Pelican Books
LIFE PULSE

Niles Eldredge is Curator and Chairman of the Department of Invertebrates at the American Museum of Natural History in New York. His interest is invertebrate paleontology, and he specializes in trilobites – extinct Paleozoic arthropod relatives of crabs, shrimp and horseshoe crabs. His major interest is the way in which the evolutionary process works from analysis of patterns of evolutionary history preserved in the fossil record. He has written many articles and several books on paleontology and on evolution for both scientific and non-scientific readers. Among his publications are *The Monkey Business* and *Time Frames*.

Life Pulse

Episodes from the Story of the Fossil Record

By
Niles Eldredge

With
Illustrations by Lisa C. Heilman Lomauro
and
Photographs by Sidney S. Horenstein

Penguin Books

PENGUIN BOOKS

Published by the Penguin Group
27 Wrights Lane, London w8 5tz, England
Viking Penguin Inc., 40 West 23rd Street, New York, New York 10010, USA
Penguin Books Australia Ltd, Ringwood, Victoria, Australia
Penguin Books Canada Ltd, 2801 John Street, Markham, Ontario, Canada l3r 1b4
Penguin Books (NZ) Ltd, 182–190 Wairau Road, Auckland 10, New Zealand

Penguin Books Ltd, Registered Offices: Harmondsworth, Middlesex, England

First published in Great Britain by Facts on File Ltd 1987
Published in Pelican Books 1989
10 9 8 7 6 5 4 3 2 1

Printed and bound in Great Britain by
Richard Clay Ltd, Bungay, Suffolk

To Michelle, Douglas, and Gregory
who have meant so much
to the story of my life

Contents

Acknowledgments

The author wishes to thank the following:

Sidney S. Horenstein, for the photographs appearing in Figures 1, 4-12, 14, 16-20, 22-23, 26-31, 34-37, 39, 41-42, 45-46, 48, 53, 57, 74, 80.

Lisa C. H. Lomauro, for the drawings prepared for Figures 2, 13, 15, 21, 24-25, 32, 40, 44, 47, 55-56, 59-60, 67-68; also for the jacket illustration, the endpaper, and the illustrations facing chapter openings.

The Department of Library Services of the American Museum of Natural History, for the photographs in Figures 3, 33, 43, 49-52, 54, 58, 61-66, 69-73, 75-78, 81-85 (Museum negative nos. 119784, 124626, 326737, 322873, 35632, 35679, 35634, 326203, 119049, 329319, 36191, 17833, 19508, 36246, 3369A, 322700, 319167, 325127, 35605, 313168, 323015, 63452, 326493, 322448, 333452, 332463, 2A333, 120963, and 310705, respectively).

Neil Landman for the photograph for Figure 38.

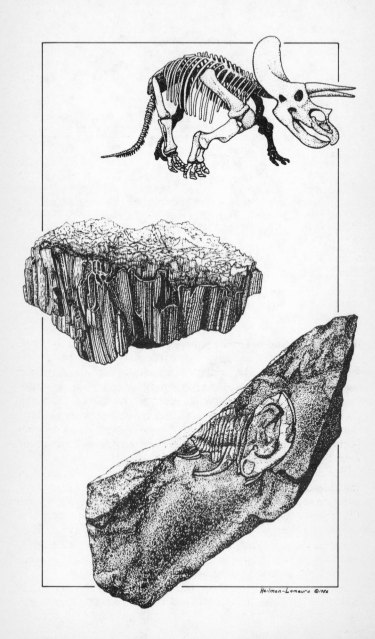

Heilman-Lomauro ©1986

I

A Walk Through Time

Visitors to the spectacular tombs of Tutankhamun and the thirty-five other New Kingdom pharaohs usually travel from the fertile Nile Valley to the arid, hidden Valley of the Kings by car or bus. The valley is carved out of a very tough, dense brownish limestone; steep hills rise up hundreds of feet to wall it in. Beyond, on the Nile Valley side, the extraordinarily beautiful mortuary temple of the pharaonic queen Hatshepsut lies nestled against imposing cliffs. The buses usually stop there on the way back to the Nile from the royal tombs.

But there is a better way to go. From Tut's tomb, several spidery tracks lead up the steep outer valley wall. A brisk climb takes you up to a narrow plateau commanding a vista of the green Nile Valley with the tan eastern desert just beyond. Closer in lie the mortuary temples of Rameses II and III, the Colossi of Memnon, and a host of other treasured ruins. And right below your feet is Hatshepsut's temple, easily reached now with a simple stroll down a gentle path. It's worth skipping the air-conditioned bus ride to get the view.

But if the view is all you get, you have missed out on another of Egypt's seemingly endless contributions to the passage of time. For in climbing that hill, you do more than leave one archeological treasure for another: you also climb through several million years of Earth history. The first time I made that climb I was anxiously scanning the rocky surface for signs of fossils—traces of ancient life. The rocks were known to have been formed from lime-rich muds deposited on the floor of an ancient forerunner of the Mediterranean Sea some 55 million years ago. I was looking for sea urchins, clams, snails, and other sorts of shelled creatures that are often preserved in such rocks. They, too, would be 55 million years old or thereabouts, vastly older than the 3,500-year-old temples and tombs of the neighborhood.

The search was fruitless. As I climbed, the only urchins I saw were the little boys who swarm around all visitors, hawking crudely faked artifacts or simply demanding *baksheesh*. (The boys come from the storied town of Qurna, right near Hatshepsut's temple; Qurna has produced tomb robbers and bilkers of tourists for over 2,000 years, starting with ancient Greek travelers who came to marvel at the ruins.) But one of the kids, noticing my preoccupation with the rocks along the path, reached into his voluminous pocket and, in a glorious, inventive mixture of German and French, produced a handful of fossils along with the words 'Muscheln, Muscheln, Monsieur!" *Muscheln* are clams, and clams these were—some nearly the size of a baseball. And all literally as hard as a rock. Genuine fossils, indeed. But where had they come from? By the time the boy had figured out what would get my attention, we had reached the top of the cliff and were experiencing that magnificent view. And all the way up there had been nary a sign of anything remotely resembling a fossil.

I offered to pay him to take us to where he had found his specimens. Easily done: there is a little hummocky hill, an erosional remnant, perched on top of the plateau—a little extra climb of about twenty-five feet that the main trail skirts around. Up we went—and there they were, littering the ground, no challenge to the eye. Clams, snails, even a crab carapace. My companions and I soon had a neat little collection—and an extra dimension to the already heady trip through time that one experiences on a visit to the archeological splendors of the Nile Valley. Thanks to this young entrepreneur, we had stumbled on a little window into time, a window that gave us a peek at what was going on in an ancient sea some 55 million years ago.

Such glimpses never fail to get me going. For life has had a history, just as human civilization has had a history. And this history can be understood. We can come to grips with the seemingly chaotic and multitudinous array of actors and forces in the drama of life's past by following relatively few and rather simple procedures. We can study

FIGURE 1   Fossils collected above Hatshepsut's mortuary temple, Western Thebes, Egypt. The four clams and one snail were culled from hundreds of specimens.

the millions of species of plants, animals and micro-organisms still alive and come away with a very real sense of history from that experience alone. We can go to the rocks and assemble as many "windows" as we can, and watch what happens—how the characters change—as we go up and down through time. For the fabric of life has remained very much the same through the last 3.5 *billion* years, even though there have been some significant changes in the composition of the world's creatures through all those years. Only an expert would perceive modifications in the sorts of clams, crabs, and snails that occurred over the last 55 million years—so trivial are those changes. Those Egyptian fossils look to the casual eye as though they could be the shells of modern clams and snails—if they weren't rocks, that is. But when the passage of time begins to be measured in the hundreds of millions, and billions, of years, changes are more easily detected. For almost 3 billion years, no life form consisted of more than a single cell; today we tend to think of "life" as nothing but multicellular animals and plants—forgetting (because we cannot see them) the millions of kinds of bacteria and more complex single-celled organisms that are every bit as numerous now as they have ever been.

Yet throughout it all, the very nature of life has always had a few fundamental aspects that were there from the beginning and have never changed. They are the very conditions, the very essence, of life. And when we look at these simple themes, we get an idea about how we can tackle the job of understanding life's history.

Life's Warp and Woof

What is "life"? Though philosophers have been wrestling with the issue for millennia, we can make some definite headway by looking at the problem in the prosaic terms of biology. Any answer will have to be true of absolutely all things we think are "alive"—a sort of least common denominator list of all life. That in itself is interesting, because such a list is likely to be short, involving the very basic facets that must have come first when life had its beginnings on earth. Thus, if we can specify the several fundamental aspects of living systems common to them all, we will more than likely also be outlining the general conditions that originally had to be met for life to have sprung from non-living matter. And once we have done that, we have provided ourselves with a key to unraveling the history of life from its inception, well over 3 billion years ago, right up to the present moment.

Fortunately, the Earth is extraordinarily conducive to "life" as we know it. We have an atmosphere with a conveniently appropriate concentration of oxygen; we have a lot of free-standing water. The sun warms us just to the right amount appropriate for growth and maintenance of complex structures—organisms' bodies—built of complex organic molecules. The other planets in the solar system have inappropriate concentrations of gases (many of them poisonous), no liquid water, and are on average either far hotter or much colder than is our Earth.

Yet for all this lucky match between our physical environment and the requirements of biological systems, the external world is still quite hostile to life. A basic necessity for any living system, regardless of size—from a bacterium to an elephant—is that it must be packaged; that is, it must have an outer defense against the elements plus internal processes that quite literally keep the home fires burning. For every organism is a factory, requiring a constant input of energy for the synthesis of new molecules such as proteins. Proteins are used for a wide variety of functions: as enzymes mediating chemical activities within cells, as building blocks for continued growth, or simply for the repair and general maintenance of the body—keeping a sort of status quo. In short, every single organism on earth is an economic machine, obtaining energy, storing it, and using it as required simply to exist: to maintain itself and perhaps to grow.

The notion of life as a matter of collections of complex carbon-based molecules capable of procuring energy and using it simply for continued existence is accurate, but not the entire story. Given the unlikeliness that any such single system, once started, will last forever (if for no other reason than that energy sources are notoriously fickle), something else seems required. And that "something else" is the ability of organisms to make more of themselves. If death is the way of all flesh, it is nonetheless true that all kinds of organisms possess the machinery for reproduction. Were that not true, "life" would have ceased long ago. And there would be no history of life for us to contemplate.

But reproduction is far more than the mere matter of keeping "life" going. Without it, there could never have been life in the first place. Nearly everyone by now has heard of the two giant molecules implicated in heredity: deoxyribonucleic acid (DNA) and ribonucleic acid (RNA). In general, DNA stores the "information" that every organism needs to carry out all those economic functions. As we grow from fertilized egg to full-sized adult, with a body composed of billions of cells of over 100 different kinds, the building instructions come from the DNA which is identically repeated in each cell. As we go about our daily lives, as our body replaces worn out cells, and as each vibrant cell maintains its own inner workings as well as performing its assigned task in the economy of the body, it is from the DNA of the nucleus and the power stations known as mitochondria that the instructions for these activities come—instructions transmitted and carried out by the various forms of RNA. These huge molecules of DNA and RNA integrate the very economic workings of the body.

The understandable fascination with DNA and RNA have led some biologists to suppose that they are the very elements of life: since their presence is necessary to integrate and regulate the economic activities of organisms, can we not suppose that in some real sense DNA and RNA are "more fundamental" to life than the economic activities, the procurement and transformation of energy? No, we cannot—for it turns out that we are dealing with a truly symbiotic system here. Just as the economic side of life is utterly dependent on the informational side to tell it how to do its job, energy is needed for DNA to make copies of itself; after all, it is the self-replication of DNA molecules that lies at the base of ongoing cellular function in organisms *and* in reproduction. Energy is required to synthesize and assemble the components of a strand of DNA, matching it up against its "parent." Then, too, at least under the sorts of atmospheric conditions we have had for the past 2 billion years, naked DNA simply cannot survive; it is bound to be oxidized, or otherwise ravaged by ambient, alien compounds. It needs a protective envelope merely to exist, let alone to provide it with the energy necessary to function.

So we have a tight circle here: in defining the twin themes that go into the very definition of life, we see that they are inseparable. Neither can exist without the other; each is utterly dependent upon the other to function. And that is why most biochemists wrestling with the problem of life's origins feel they must solve two problems simultaneously: the origin of proteins, and the origin of DNA/RNA.

Economics, Information, and Life's History

Why should we be overly concerned with the chemical events of billions of years ago? After all, the origin of life is hardly a subject of direct paleontological concern. We have no fossil specimens of primitive DNA strands wrapped in a skimpy little protein blanket. The oldest fossils we have, fossils some 3.5 billion years old, while hardly higher life forms, are nonetheless full-fledged bacteria, every bit as complex as the bacteria we have with us today. The fossil record simply isn't much help in investigating life's very beginnings. The problem lies squarely in the chemistry lab.

But in listing life's essential ingredients, and thereby stumbling upon the question of life's origins, we have also gained some insight into what we might expect as we survey the sequence of events in life's long history. For in a very real sense, the history of life is the story of the maintenance and modification of the economic side of life; and major evolutionary change typically involves new inventions (such as bird's wings, or multicellular bodies), nearly all of which are concerned with the procurement and transformation of energy for the ongoing existence of individual organisms. (Some inventions pertain purely to reproduction, such as elaborate courtship behaviors, etc.) Indeed evolutionary theory to the present day sees the evolutionary process almost strictly as a matter of the modification of the genetic information that underlies features of organisms that we call "adaptations." Darwin's mechanism of "natural selection" maintains or modifies adaptations as they affect the survival, and especially the reproductive success, of individual organisms. Since organisms are both economic machines *and* reproducers, the better economic machines will leave disproportionately more of *their* genetic instructions to succeeding generations than will their less economically viable peers—in a world so harsh that only a fraction of the organisms of each generation can possibly survive and reproduce. We live in such a world.

Thus it is traditional to look at the history of life as a story of the progressive modification of adaptations. And the fossil record superficially bears out such a view: all we see in 3-billion-year-old rocks are bacteria. Today we have bacteria plus a lot of other, far more complex organisms. But the road to progress, if such it be, has been bumpy. Yes, there is certainly the theme of simple progress in the ongoing construction of better mousetraps (i.e., more efficient organic economic machines); but a far more dominant theme is that whatever system is up and running at the moment—whatever *works* economically—will be conserved. Little tinkering goes on. Evolution, or so I firmly believe the fossil record tells us, is no matter of inevitable improvement in economic efficiency, no story of constant competition to find yet another superior modification to build a slightly better mousetrap. It is no race to invent wholly new, improved economic devices.

Instead, evolution is much more a matter of survival, its first rule being "If it works, don't mess with it." Indeed, if it were not for those bumps in the rocky road of life's long history, bumps sometimes so severe that extinction claimed huge proportions of all species alive at the time, there most likely would not have been anything near the amount of change that we can in fact see as we scan life's 3.5-billion-year-long fossil record. Thus, by examining the fossil record more closely, we get a rather different picture of life's evolutionary history than we might by considering only natural selection as it works to maintain and modify the economic adaptations of organisms.

Genealogy, Ecology, and the Fossil Record

Darwin knew that his central view of evolutionary change—slow, steady, progressive, and *adaptive* modification—was at odds with what was then known of the fossil record of life's history. He was thus forced to lay the groundwork of yet another science—*taphonomy*, the study of the fossilization process—to explain just why the fossil record is so poor, riddled with gaps of missing information that give life's history the *appearance* of periodic change.

Nowadays it has become fashionable to take the fossil record more literally. A century of experience since Darwin's *On the Origin of*

Species (1859) appeared has convinced many paleontologists that there really are gaps in the record, but that many of those gaps are actually *caused* by events in the evolution of life. There seems to be something fundamentally episodic about the events that have shaped life's history.

And when we confront these patterns of stability interrupted on occasion by events of evolutionary change, we immediately discover the twin themes of economics and information that have been there ever since life's beginning. For there are two ways of looking at life's history: we can trace up the separate branches in strict genealogical fashion, or we can go sequentially up the geological column and look at absolutely everything alive at any one moment (so far, that is, as they are preserved and we can find them).

Most professional paleontologists think of evolution in genealogical terms. Each of us has one or two groups of organisms with which we are especially familiar—large arrays of related species that arose at some point in the recesses of geologic time, persisted for some amount of time (usually reckoned in the tens of millions of years), and perhaps already extinct. My own group, for example, are the trilobites (extinct, primitive relatives of horseshoe crabs and crustaceans). To be even more exact, I specialize in a particular group of large-eyed *phacopid* trilobites, which flourished over an interval of some 150 million years, from the so-called "Lower Ordovician" Period through the "Upper Devonian" Period. (These are subdivisions of geologic time, shortly to be discussed more fully.)

Yet it is apparent to all paleontologists that organisms have *two* sets of relationships. In terms of genealogy, we humans, for example, are close relatives of the great apes and certain fossil forms of proto-humanity; more remotely, we are related to monkeys and various "lower" primates, and so on; the skein of life interlinks absolutely all creatures that have ever lived on Earth.

But equally important are our ecological (or, simply, economic) relations: all organisms live in local ecosystems, networks in which energy is being transferred in complicated yet fairly orderly ways. These ecosystems are the actual arenas of life—where the actors are playing out their roles as far as evolution has taken each of their lineages. Ecosystems are functioning, vibrating manifestations of the fruits of the evolutionary process. Viewed this way, species and larger collections of evolutionarily related species (*taxa*) seem like ledger books of the genetic information that allows organisms to succeed in that other realm—the economic world.

And perhaps not surprisingly, this economic perspective suggests another way of looking at the history of life—an alternative to specializing in the vicissitudes of a single genealogical group, such as the trilobites, as they appear, live on, and eventually become extinct. Instead we can go to the rock record and trace the histories of entire ecosystems, economically interacting assemblages of organisms of a wide variety of pedigrees, all living together cheek by jowl in some ancient seabed, marsh, or arid plain.

It is because I am convinced that an accurate grasp of the evolutionary process can only come from an appreciation of how economic systems interact with genealogical systems that I have chosen to follow both paths in this book. Following my own tradition, I discuss coherent, evolutionary groups—such as trilobites, snails, dinosaurs—tracing their histories and presenting insights they may have shed on the general nature of life's evolutionary past. These groups—organisms, species, and larger entities—are, after all, the actors on the evolutionary stage. We need a program to tell who the players are.

Yet it is easier, I think, to get a sense of history if we trace life, or as much of it as we can embrace in these pages, in the single sequence of time and over the changing surface of the Earth—the stage on which all this drama has been enacted. That is why this book starts at the beginning of earthly and biologic time and moves on in linear fashion to modern times. The one departure from this organization—Chapter 7, which deals with the phenomenon of extinction—focuses on the most famous of all mass extinctions, the event that terminated the Mesozoic Era, taking out the dinosaurs and many other, less celebrated creatures; that chapter follows, appropriately enough, a look at life in the Mesozoic (Chapter 6).

Evolutionary Patterns in Life's History

What, then, is the pulse of life? If Darwin's expectation of gradual, progressive change does not emerge, what do we really see as the main signal in life's history? I have already said that the patterns are episodic in nature, and that evolution has a tendency *not* to tinker with systems that are already up and running, and functioning very well. Evolutionary change is far from the inevitability that, given environmental change over the passage of time, many have expected it to be.

The episodic nature of life's history presents an interesting theme, because this phenomenon of business-as-usual, stable systems undergoing change at rare intervals, goes on at several *scales*. In local systems, such as the isolated woodlot still to be found in many a suburban setting, the local representatives of a wide variety of plant and animal species are by no means permanent: some are transients even though others live their entire lives in that setting. In some years, certain species (such as migratory birds) will not be in the lot at all, while other species will have representatives there continuously for hundreds of years. Yet the cycle of birth and death will give even the permanent residents at least an aura of episodic comings and goings. At this micro-scale the fossil record will reveal nothing; yet naturalists focusing on the details of microevolution—where what happens to each generation of organisms of each species is very much of interest—need to take the episodic nature of the life of this woodlot ecosystem into account.

Moving up a notch we see that local ecosystrems are parts of larger systems—such as large sweeps of the ocean floor with its fairly homogeneous array of vertebrate, invertebrate, and microorganismic life. We often find entire species confined to such large arrays. And paleontologists have in recent years become impressed with patterns in the comings and goings of entire species, for, it turns out, species have births, histories, and deaths much as do organisms. The theory of punctuated equilibria, which I established with my colleague Stephen Jay Gould,* attempts to explain the episodic nature of the comings and goings of species: for species tend to appear rather abruptly in the rock record, frequently lingering for millions of years with little, if any change—and then abruptly disappearing.

As we will see as we go through life's history, ecosystems themselves reflect the sort of stability we typically find in individual species. Ecosystems in the marine realm will remain stable for 5 or 10 million years. To be sure, over the course of such a vast amount of time, some species will disappear, while others will appear (whether newly evolved or simply introduced from elsewhere). But nothing *much* happens either to the overall complexion of the ecosystem or to the players in the game—the species that supply the organisms to fill those ecological roles—throughout the long, "normal" phase of an ecosystem's history.

The Importance of Extinction

How then do we get any *change* in the history of life? The answer, increasingly, seems to lie in what has always seemed to me to be the

*See my book *Time Frames* (New York: Simon & Schuster, 1984) for a full account of this theory and its history.

very antithesis of evolution: extinction. We speak of the demise of a species as its extinction, but if we look at the ecological side of things, we see that the history of life is absolutely riddled with extinction events that affect entire ecosystems, accounting for the demise of some varying percentage of all different sorts of organisms comprising the system. If a woodlot burns, all but a few microorganisms will disappear. In such local situations, of course, entire species do not become extinct.

But consider larger ecosystems. Large-scale degradation of regional ecosystems could well spell disaster for many unrelated species. And that is perhaps the loudest message that the fossil record has for us: large-sccale, cross-genealogical extinction events, brought on by the wholesale degradation of large ecosystems, occur regularly in the fossil record. To date, of course, no single event has succeeded in wiping out everything. But some extinctions have been so massive, so nearly worldwide and pan-habitat in nature, that they have wiped out as many as 95 percent of all living species.

I'll have more to say about extinctions as the book unfolds. But consider what happens *after* an extinction event: there is vacant ecological space. That is when evolution becomes its most creative, rapidly (usually following a characteristic lag) producing a wide variety of creatures to take the places of their fallen comrades. I firmly believe that without extinction to free up these ecological niches, life would still be confined to a primitive state somewhere on the sea bottom.

And here we have touched on the very pulse of life, and gotten very close to an understanding of how economic affairs of animals and plants relate to the comings and goings, the very evolutionary histories, of entire groups. The severity of an extinction profoundly affects the histories of whatever species inherit the Earth on the next go-round of proliferation. Minor extinctions, for example, typically leave some species that are closely related to those that failed to survive; as a rule, such close relatives and their descendants are the ones that reoccupy the terrain once life recovers.

But larger-scale extinctions—where entire families and even larger groups disappear once and for all—often leave no truly close relatives to take over. Other, less closely related and less ecologically similar organisms then get a crack at the job—this was the situation, apparently, with the mammals, who only got a chance to "make something of themselves" once the dinosaurs had finally, and irrevocably, departed this Earth. I am speaking metaphorically, of course: organisms in no sense "choose" to evolve. Yet the exploitation of niches left by extinct organisms is perhaps the most profound theme of the entire history of life.

Thus this book, with its dual preoccupation with genealogy and ecology, with the extraordinary range of organisms and sweep of time in life's history, is of necessity itself somewhat episodic. I have dwelt on those creatures, those ecosystems, and those particular events that strike me as particularly illuminating of the overall themes of the pulse of life through time. I have tried, as well, to convey some flavor of the experiences working paleontologists have had as they unraveled the past. After a brief look at some of the problems in dealing with the vastness of large biological entities and of geological time, we will be in a position to begin at the beginning and seek to get a reading on the pulse of life's past.

Seeing Big Things

We were driving through Badlands National Monument many years back. As the car dipped down a small hill, passing layer upon layer of pink and gray silt and clay beds, the driver (who was not a paleontologist) casually asked me how much time we had just driven through. I had no exact idea, but I was struck with his thoughtfulness in simply asking the question; I told him we had just descended perhaps as much as a million years.

Time is rarely evenly recorded in sedimentary rocks. The layers of the Earth's crust that preserve the fossil record are built up by accumulations of muds, sands, gravels, and limes that keep piling up as the ages roll by: there really is a simple relationship between the thickness of a pile of sediment and the amount of time it took for it to accumulate. But that's only part of the story. We may be tempted to speak of the gradual accrual of muds over a period of 500,000 years, culminating in a fifty-foot-thick body, rightly pointing out that fifty feet of mud is exceedingly unlikely to accumulate overnight (though it may in extraordinary circumstances, such as a landslide). We might be further tempted to calculate an average rate of deposition: if it took half a million years to accumulate fifty feet of muds, that breaks down to a foot every 10,000 years, or a little over an inch every thousand years.

Yet such numbers are apt to be grossly misleading. For deposition of muds is far more an episodic, now-and-again affair than the simple calculation of average rates might lead you to believe. Not only can storms, earthquakes, and other sorts of periodic natural disturbances trigger the sudden deposition of relatively large amounts of sediment, but the opposite can happen as well: often the source of all that sediment is simply shut off, and *no* muds or silts are deposited for thousands of years. For example, in the vast seas that covered the continental interiors for much of life's history, the bottom muds we now comb for fossils came from two basic sources: the limy components are derived largely from the shells of the organisms actually living there, while the inorganic muds (mostly particles of clay minerals and quartz) were derived from the land fringing these inland seas. Particularly near the margins of these vast stretches of water, the episodic availability of muds and silts washing in from the land is dramatic. Along the eastern side of the inland seas you can actually trace pulses of mountain-building as the proto-Appalachian chain rose spasmodically. One such phase is clearly recorded in New York state, in rocks between 380 and 360 million years old. In those 20 million years, a great delta slowly spread from east to west across what is now New York. But it did so in bursts. In any one spot, each pulse of mountain building is recorded as a relatively thick wedge of fairly coarse-grained sediments—a simple reflection of streams rapidly coursing down the freshly thrust-up hillsides, charged with fresh loads of eroded soils and rock. As the hills wear away, the streams slow down, the sediment load drops, and the grains become finer as they are no longer masked by the larger grain sizes. Thick deposits of sands and silts are succeeded by thinner accumulations of fine-grained muds. Yet the overall direction of change was the uplift of a substantial mountain chain—so the net result was an encroachment of land over sea, with the delta preceding it, from east to west.

Any body of sedimentary rock represents the passage of some chunk of geological time. But except in extraordinary circumstances, there is no way of knowing how much, or which parts, of that time is actually represented by rock—and how much of that time is missing as gaps in the rock record. It is simply in the nature of things for there to be gaps in the rock—hence the fossil—record. And this state of affairs has traditionally been supposed to deal a knockout punch to paleontology as a source of detailed inference on the nature of life's evolution. We might be able to chart the gross outlines of life's actual history on Earth, but especially when it comes to determining *how* life has evolved, we are widely supposed to be plumb out of luck. After all, evolutionary theory ever since Darwin has focused on natural selection as the prime mechanism of change, and to study natural selection we must see patterns of change in gene frequencies (or genetically based properties) within populations—all this on a generation-by-generation basis. And no one disputes the basic implausibility of doing any such thing with fossils.

I have already mentioned that it was Darwin who, more than anyone else, brought the inadequacies of the geological and paleontological record to our collective attention. He had to: he felt he had to explain why there weren't far more examples of slow, gradual evolution in the fossil record—examples that would, if only they existed, abundantly confirm his theory. But there has been a very positive fall-out from Darwin's discussion of gaps in the rock record: he essentially founded an entire area of paleontological concern, in which we seek to know how creatures come to be fossilized in the first place, thus how much of the actual history of organisms we can realistically expect to sample as we explore the fossil record.

But all this talk of gaps and inadequacies remains something of an embarrassment to the profession. It will never do simply to make excuses for your data: the impression inevitably created is "Well, I realize the data are horribly deficient, but if we make certain allowances (and perhaps close our eyes and wish real hard), we can nonetheless move forward. . . ." An alternative is to focus on what is *good* about the fossil record—emphasizing some of the remarkably complete and extraordinarily rich circumstances occasionally encountered in the fossil record. This at least accentuates the positive.

Yet I prefer a third way of looking at the record. Yes, the fossil record is incomplete in many ways, for many reasons. Add limitations of access and actual exploration and study to the inherent limitations of preservation. And yes, despite this incompleteness, there are, as we shall see, some deservedly celebrated instances of amazing wealth in a few nooks and crannies of the fossil record. But I see a virtue in the record's incompleteness, the very feature so commonly held to be the record's greatest flaw ever since Darwin. For where else, *how* else, can one sample the *history of an entire species*? Suppose you drove through Badlands National Monument, up hill and down dale, traversing time as you went through space. The stream and lake-deposited sediments there are mostly Oligocene in age—and altogether some 20 million years are (sporadically!) preserved—an interval of time beginnning approximately 37 million years ago. The layers reveal an abundance of fossils—mostly bones and teeth of mammals—weathering out of the sides of the buttes and concentrated in erosional gullies. The floodplain ecosystem of Middle America 30 million years ago comes through pretty clearly. We see herbivores, such as the extremely common oreodonts, extinct and rather sheep-like (or even pig-like) relatives of, well, sheep; carnivores (a wealth of dog-like, cat-like, and "bear-dog" forms). We even know something about the grasses and other plants that formed the base of the local food chain: pollen grains are easily preserved, and each species, like human fingerprints, has its own distinctive set of swirls and sculpture.

But let's just take one species—say a species of the common oreodont genus *Merycoidodon*, easily collected in the Badlands. True, because it did not cover every square inch of turf for every minute during our Oligocene interval, and because of the vagaries of deposition and bone preservation (most bones are thoroughly destroyed before they have a chance of getting buried), it is hopeless to expect a complete record of this species. In any case, we would be faced with countless *millions* of oreodonts, thus *billions* of bones, if the record were complete.

So what, then, do we have? We have some samples, the earliest known, of specimens that appear to belong to this species of *Merycoidodon*. We can safely assume these earliest known specimens do not record the very beginnings of the species, though there is a remote chance that they might. In any case, we then have a serendipitous sampling upward through time—the occasional specimen, the odd bonanza where many specimens come from one crowded jumble of bones from a relatively restricted horizon. And inevitably, we have the highest (latest in time) known sample—which again may or may not represent the last individual of that species ever to exist. In all likelihood the last known fossil is really the remains of an animal that died thousands, if not hundreds of thousands or even millions of years before the entire species became extinct. Yes, the sampling is by far too coarse to assess the generation-by-generation workings of natural selection. But in the place of such refined information we have something else—something we can learn nowhere else.

For we know what happened to that species *in toto* nearly throughout its entire existence. We have a collective snapshot of the entire history of that single species—or rather, a series of snapshots similar to the photographic history of any of us in a technologically advanced society. Few of us are photographed every day; but the odd birthday, anniversary, Thanksgiving—assemble them all and there is a more or less haphazard, gap-ridden photographic history of a single entity—one's self. The analogy goes still further: contemporary practices in paternity aside, few of us are photographed just at birth. We are usually a day or two old before the hospital photographer first captures our image. Nor are we usually photographed on our deathbeds. Yet withal, our photographic histories are nonetheless valuable and useful documentaries of the passage through our lifetimes.

Think back, not of individual organisms such as yourself, but of entire species. There are two ways to go here, and which way you choose makes a tremendous difference in how you will think about the significance of the entire 3.5-billion-year fossil record of the history of life. We could play it safe and take the conventional view: yes, we have a "photographic album" of sorts, an erratic snapshot history of the bulk of a particular species' time on earth. Though this record is too coarse to document the precise effects of natural selection, it nonetheless allows us to chart the changes (or lack thereof) in a gross sort of way, which perhaps can set up a gross sort of constraint on how we might imagine the evolutionary process—the process that gives us life's history—to work.

And that is precisely the conventional view. The fossil record imposes some few broad limits on how fast or slow natural selection can work to effect change. The plain truth is that most species hardly change to any noticeable extent over their histories—histories that typically last 5 or 10 million years, or even longer. The fossil record has yielded pretty well established facts about the typical rates of organic change through time.

But, as I keep hinting, there is an even more profound way to interpret such information. We can push the "snapshot" analogy still further—without, I think, stretching its limits or straining credulity. For why can't we think of an entire *species* as in some profound sense entirely analogous to an individual organism—you, or me, in our separate photographic histories? I do not want to push this too far: I am not, for one thing, proposing that species age in a manner similar to organisms. But I am saying this: species, like organisms, have histories. And they have births and deaths. In other words we can think of species as *units*, *"individuals"* if you will, just as organisms are "individuals."

Thus the gappiness of the record artificially "telescopes" large-scale biological entities. Vast stretches of time are represented in manageable piles of sedimentary rock. We can study not only the entire comings and goings of species, but also the vicissitudes of larger groups (such as dinosaurs or trilobites). And returning to economic matters, we can also see ecosystems of various sizes as historical entities—watch them become established, see them become modified through time, and witness their ultimate demise—the better to understand the processes underlying life's history. That gappiness of the record, far from being a plague that removes paleontology from any meaningful position in evolutionary biology, is the *only* way that we can study the histories of large scale biological entities.

Time in Space

North America is blessed with many natural resources, and as a paleontologist I am glad to say this certainly includes fossils. Of all the beautiful spots to collect an exciting array of specimens, my personal favorite is *not* in the spectacular Rockies, or even along the charming coast of the Gaspé Peninsula (in Canada's Quebec Province); rather, it is in the American Midwest. Indiana, to be precise. Part of this perhaps odd-sounding choice simply means that *my* kind of fossils—trilobites of a certain age and identity—are commonly and well-preserved in the quarries and roadcuts of that state. But this accounts for only part of my paleontological fascination with Indiana: for trilobites or no, Indiana exposes over 100 million years of time in a series of north-south strips that run pretty much the length of the state. Start in the east—in Madison, say, down in the southeast corner, on the north bank of the Ohio River—and simply head west. Crossing Indiana's midriff, not only will you climb that 100 million years in time, you will also (if you know where to go) see some of the best-looking fossils you could ever hope to find.

We'll have occasion to look at some of these Indiana fossils and ancient ecological settings a bit later, when we catch up to the Earth and its life beginning about 450 million years ago. Right now it is the whole package of Indiana space/time that invites exploration. For I have bandied about items like "450 million years" as though such figures were intuitively obvious in both scope and origin. I have used terms like "Oligocene" and "Devonian" as if they were household words. And though I have said that sediments, sensibly enough, fill up basins (like lake and seabeds), so that the layers at the bottom are formed earliest, the upper ones of course coming later, I am now telling you that a drive *across* Indiana is a drive through time, up or down, depending on how you go. But this is not like those dips and rises within Badlands National Monument: Indiana is far too flat to expose 100 million years of time vertically. So we need to know a few more things about rocks and time before proceeding with the fossilized history of life.

It was Darwin, more than anyone else, who gave us a firm *intellectual* sense of time—vast hunks of it. Darwin's early training and experience was in geology. It stood him in good stead: for to make his notion of evolution be seen as workable, Darwin required far more than the paltry millions of years which scientists of the mid 1800s were willing to grant as an age of the Earth. Darwin drew on his geological expertise as he talked of the sheer enormity of time it must take simply to erode cliff faces and deposit huge thicknesses of sediments. It turns out that he exaggerated: the Grand Canyon (which Darwin never discussed) is the greatest erosional rent on the subaerial surface of the Earth, and even there it apparently has only taken about the past 1 or 2 million years for the Colorado River to do most of the cutting down through that mile or more of rock. But Darwin had the right idea about the general order of magnitude of geological time—just as he proved so basically right about the evolution of life.

Now we know that the Earth is 4.55 billion years old (give or take a few million years). When I say "we know," I mean that three in-dependent means of getting a *number in years* for the age of the Earth consistently converge on that figure. (The three sources: age of the oldest moon rocks; age of stony meteorites that fall to Earth; projected age extrapolating back from the oldest rocks yet dated on the Earth's surface, which can be sited on a continuum from youngest to oldest—4.0 billion years, at last report.) Nothing being absolute in science (including the notion that the Earth is an oblate spheroid), what I really mean is that repeated experience points to that figure. It is extremely well established. It is as close a thing to "fact" as we can reasonably expect in this world.

Where do we get these numbers, these fairly exact estimates of the age of rocks? We get them from knowing the statistical rates of decay of radioactive isotopes of various elements—uranium and thorium for very old rocks, potassium for not-so-old rocks, carbon for younger objects. A problem immediately enters, for sedimentary rocks—the hardened muds, clays, sands, and limes of lakes and seas—generally do not contain any freshly formed minerals that will start "ticking" like radioactive clocks the moment the sediments are deposited. These clocks occur in igneous rocks—rocks cooled and congealed from a liquid melt, such as granite and lavas of various sorts. But here we are in luck: granites and lavas are occasionally extruded into sedi-mentary rocks at about the time (or soon after) the sediments begin accumulating. Through those igneous rocks we can get a pretty accurate idea of when the sedimentary layers were accumulating. For example, the Palisades cliff face, along the west bank of the Hudson River in New Jersey, is a lava-like igenous intrusion that was squeezed between layers of sediments that (the fossils indicate) were being deposited near the end of the Triassic Period. You can tell that the palisades diorite (as the cooled rock is called) was injected be-cause the sediments both above and below it are altered by heat, literally baked (as you can readily see on top by looking at the rocks at the contact on the north side of Routes 180 and 95 in Fort Lee). But further west, three lava flows (the Watchung Mountains) were formed at about the same time, and we can tell that these were actual lava flows on the Earth's surface because only the sediments on the bottom of the lava are scorched: those above are not, evidence that they were deposited after the lava had cooled. All the radiochemical dates for these igneous rocks are about the same: some 200 million years ago. This accidental juxtaposition of igneous and sedimentary rocks gives us the means to put numbers on ages of fossils.

But there is another way to do it: we say, "This trilobite is Middle Devonian," and when we compare it with, say, an "Upper Silurian" trilobite, we know we are comparing it with something that is older. We do not know (without consulting a chart, such as the one presented in the endpapers of this book) *how* much older "Upper Silurian" is relative to "Middle Devonian." But the fact that it *is* older is a matter of definition.

All this comes from the historical recognition of the geological column, which now has a time scale attached. Niels Stensen (a Danish physician writing under the latinized name Nicolas Steno in the 1600s) saw that many rocks were sedimentary—simply the hardened, cemented water- or air-borne grains that had finally come to rest somewhere. It was he who pointed out the historical implications of the "first-to-be-laid-down-is-the-oldest" nature of sedimentation. By the late eighteenth and early nineteenth centuries, geologists were following sedimentary rock sequences up and down, concluding that there was a regular, consistent sequence to them—at least in any one area. The "Devonian" Period is simply named after the English county of Devon and refers to rocks with a characteristic suite of fossils. The rocks originally designated "Devonian" (by Adam Sedgwick and Roderick Impy Murchison in 1839)underlie what the English had been calling the "Carboniferous." They overlie what Murchison had named "Silurian" rocks in neighboring Wales. (The Silures were a primordial tribe living in the Welsh region.) Thus names were attached to large bodies of sedimentary rock strata that were observed empirically to overlie some other rocks and underlie still others. The sequence of names in the geological time chart reflects the fruits of these labors. Still being refined, it was worked out in its basic form by the late nineteenth century.

But that's not all. William Smith, an English surveyor helping to lay out the extensive system of canals built at the turn of the eighteenth century, noticed that fossils always occur in the same relative sequence on neighboring hillsides. Soon, when the early stratigraphic (literally, the "writing of strata") work spread from England and continental Europe, it was realized that fossils the world over occur in the same general stratigraphic sequence: fossils easily recognizable as "Silurian" in North America (simply because of the near identity with Silurian fossils from Great Britain) *always* lay above "Ordovician" fossils (so considered in North America again because of their near identity with so-called Ordovician fossils in Great Britain). And Silurian fossils always underlay "Devonian" fossils (the logic being the same) wherever they turned up over the Earth's surface.

Furthermore, enough radiometric dating analyses have been performed by now that the dates, the actual years, that I have included in the time scale seem quite accurate. We need expect no major changes: a laboratory might collect rock samples around the world, samples predicted to be "Lower Devonian" in age because the igneous rock cuts through Lower Devonian rocks but not through the overlying Middle Devonian rocks. Such labs always measure approximately the same age for rocks predicted to be about the same age. Other labs will take samples from the same rock masses and get precisely the same ages. There are even alternative chemical pathways to evaluate rock ages, and these are mutually consistent. The geological time scale, such as the version I include here, seems pretty secure, indeed.

But none of this tells us why a trip across Indiana is a trip through time. The reason is that although rocks may be formed as horizontal layers, with the youngest on top, they need not remain that way. Especially in mountain systems, but also in relatively undisturbed areas such as the American Midwest, rocks can be tilted. And it

FIGURE 2   When originally flat-lying sedimentary rocks are folded and erosion wears through the layers, a spectrum of ages is exposed on the Earth's surface.

turns out that even a gentle tilting can have profound effects on what rocks come to the surface to be exposed and picked over by fossil collectors, and which ones remain deeply buried and forever beyond the ken of paleontologists. It turns out that Indiana is nestled alongside the "Cincinnati Arch"—a positive feature of the North American crust that has been sticking up for hundreds of millions of years. No one knows why this feature is there; it just is, and it tilts the rocks so that the oldest sediments of the Midwest, which are Middle Ordovician in age, are exposed nearest to the center of the arch. Travel east to west from the center, and you will have to go through progressively younger rocks. It's that simple—an effect that can be duplicated simply by folding a deck of cards and slicing off the top of the fold.

Yet there are profound consequences of this little piece of crustal geometry. For thick and relatively undisturbed sequences such as we see in the Grand Canyon are truly rare; few places remain downwarped sufficiently long to receive such massive amounts of sands, silts, and muds. And such basins, when they do form, tend to remain beneath the ocean surface on the continental margins. Or they are swept up and distorted in the crinkling of fresh mountain chains, or sucked down and obliterated in the very bowels of the Earth along the intersection of two massive plates of the Earth's crust. So it is just as well that Mother Nature has done the job for us in another way: flat-lying strata, tilted sometimes ever so gently, will expose a potentially vast expanse of rock—hence time. And when this happens, it enables us to time-travel if we stay relatively horizontal and simply move over the terrain. Move, but something else: we must observe. For now we have all the conceptual tools we need. Now we must roll up our sleeves and be willing to travel around the globe in the present time frame: the past, the glorious variegated history of life, stands ready to be revealed.

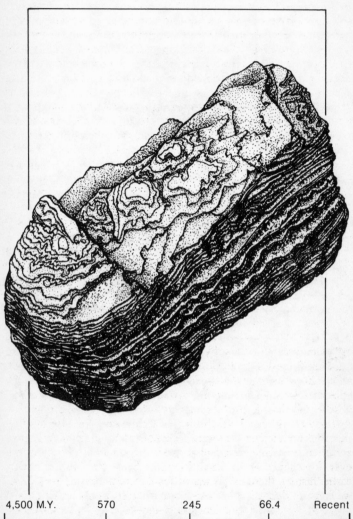

4,500 M.Y.   570   245   66.4   Recent

PRECAMBRIAN | PALEOZOIC | MESOZOIC | CENOZOIC

Heilman-Lomauro © 1986

2

In The Beginning

Science can be a humbling venture. From its earliest days as a full-fledged discipline, paleontology offered a few basic facts about life's history that seemed incontrovertible. And of all paleontological generalizations, perhaps none was so dramatic as the observation that, from the standpoint of fossils, geologic time had two major divisions. It was the universal experience of the fraternity that the very ancient rocks were devoid of animal and plant remains, and then, very suddenly, and at about the same horizon the world over, life showed up in the rocks with a bang. For most of Earth's earlier history, there simply was no fossil record. Only recently have we come to discover otherwise: life is virtually as old as the planet itself, and even the most ancient sedimentary rocks have yielded fossilized remains of primitive forms of life.

The sudden and great proliferation of complex forms of sea-dwelling animal life came at the base of the Cambrian Period (now known to be about 575 million years ago); as we shall see later in this chapter, this event remains one of the most fascinating episodes in the history of life. When paleontologists thought that no fossils were to be found in any older rocks, they did not leap to the conclusion that life had all of a sudden been invented at the beginning of Cambrian times. Particularly because Darwin had so convincingly argued tht life evolves slowly, requiring huge amounts of time to accumulate significant change, few paleontologists of the nineteenth and early twentieth centuries were willing to claim that the apparently sudden advent of complex creatures 575 million years ago was an actual evolutionary event. They preferred to see it as an artifact of geological processes: they noted that most older rocks were either igneous or the metamorphosed remnants of sedimentary rocks whose fossils had been baked and squeezed out of existence. Life *had* been there in those earlier eons, they felt sure, but simply had not survived the ravages of time. Especially since life had diversified into a wondrous array of mollusks (snails, clams, and the like), arthropods (trilobites plus more modern groups such as insects and crustaceans), echinoderms (the starfish/sea urchin clan), brachiopods (the dominant shellfish group of ancient seas), and other, less-well-known creatures, paleontologists quite naturally felt that there must have been a truly long period of slow evolution to allow all these different forms to derive from the common ancestor they were supposed to have shared.

Wrong on both counts. Careful paleontological detective work begun in the 1950s has revealed an extensive, if elusive, early fossil record. And this new Precambrian paleontology has made us take that early Cambrian event much more seriously, for it does not bear out the predicted long, slow history of diversification of complex life. That early Cambrian spurt of life looms now as one of the most important ecological and genealogical events in the entire history of life. And the events leading up to it, during life's first 3 billion years, give little inkling of what was to follow. The stage was set in those first 3 billion years, but only in the most general sort of way: looking at that unbelievably long and seemingly almost uneventful early history of life, there was simply no way anyone could have anticipated what happened so relatively quickly when complex animal life finally appeared on the scene.

Early Mists of Time

Life's *biggest* event, of course, was its invention, its very beginnings. We have already seen (in Chapter 1) that the rocks contain little direct information on this event—and in any case the problem of explaining the origin of life lies squarely in the realm of biochemistry. Yet a few salient features of early Earth history do shed some light on the problem.

Planetologists no longer believe the Earth had a fiery origin, requiring eons of time to cool down and produce a hardened outer crust. That idea was discarded once it was realized that all the heat flowing out from the Earth's interior (still palpable in mine shafts and manifested in hot springs and volcanic activity) is the product of radioactive decay deep within the Earth. No longer need we invoke molten origins to understand all that heat. Thus we could expect there to have been an outer crust surrounded by an atmospheric envelope fairly early on.

The Scottish physician and farmer James Hutton (1726-1797), effectively the founder of modern geology, once observed that the Earth reveals "no vestige of a beginning, no prospect of an end." He was impressed by the continual recycling of crustal materials. Sedimentary rocks are the accumulated debris derived from older rocks. Nothing lasts forever, and Hutton saw how unlikely it was that the very oldest rocks on the globe would survive the destructive forces emanating from the Earth's interior or eating at them from the atmosphere. He was right, of course, yet even Hutton would be surprised to learn that the oldest rocks thus far dated (Precambrian granites from Australia) are some 4 billion years old. We have closed to within half a billion years of the very origin of the planet Earth.

But as impressive as this vast age for the oldest rock might seem, it is even more astonishing that the very oldest sedimentary rocks so far discovered have fossils in them. These rocks, interbedded with lavas (thus chemically dated), are in the Warrawoona Group in Northern Australia. They are some 3.5 billion years old. Looking at these incredibly old rocks and fossils, you get the feeling that life is an intrinsic part of the Earth, not merely some latecomer, an appendage haphazardly stuck on as an afterthought of creation.

All these early forms are simple bacteria—microscopic rods and spheres. Bacteria are the simplest forms of life known on Earth, today or in the past. Thus the fossil record bears out a prediction we would make about life's history just by surveying life as we observe it today. We see that bacteria are indeed the least complex forms of life (viruses are parasites, and must invade foreign cells to reproduce; so viruses must be degenerate life forms that appeared *after* more complex cells had evolved). So we would expect that bacteria would show up first in the fossil record. And they do.

It is, of course, extraordinarily difficult to find fossil bacteria. They are tiny and extremely fragile; the first discoveries of Precambrian bacteria were in cherts, which are sedimentary deposits of fine-grained quartz (SiO_2). To see them at all, paleontologists must cut extremely thin slices of this very hard rock—slices thin enough to transmit light. Under the high magnification of a compound microscope some of these slices, derived from such rocks as the Gunflint Chert of Ontario, or the Fig Tree and Bulawayo rocks of South Africa and Zimbabwe, respectively, began to reveal traces of bacteria that some paleontologists (notably Stanley Tyler of Wisconsin and Elso Barghoorn of Yale) simply knew *must* be there.

Bacteria are classified as *prokaryotes*. The most basic subdivision of life is simply the distinction drawn between these "prokaryotes" and a coordinate group, the *eukaryotes*. You can think of them as the "haves" and the "have-nots"—eukaryotes possess a complex cellular anatomy, with a double-walled membrane separating the nucleus from the rest of the cell's interior, plus a host of other refinements. Prokaryotes—the "have-nots"—simply lack such refinements. All prokaryotes are bacteria of diverse sorts; eukaryotes are everything else: amoebae and other sorts of single-celled and generally microscopic creatures, as well as fungi, plants, and animals—including, of course, ourselves.

For about 2 billion years life consisted solely of prokaryotes. The second big event, which occurred no later than 1.3 billion years ago (the age of the earliest eukaryotic fossils so far discovered) was the invention of the more complex, sophisticated cell type. And again, little can be inferred directly from the fossil record about the nature of this event. Biologist Lynn Margulis of Boston University has mustered considerable evidence from the comparative anatomy of living forms to support her notions of the hybrid origins of the eukaryotic cell. Her idea is that several different kinds of prokaryotic organisms literally joined forces, dividing up the labor as some of them became specialized "organelles"—discrete parts of a cell, each with a particular function. Several of the organelles—such as the mitochondria of animal cells that function as the cells' power plants—have their own DNA. Margulis believes that several sorts of primitive prokaryotic cells invaded the cytoplasm of still other cells, creating a symbiotic system. Thus eukaryotes are permanent hybrids of several forms of prokaryotes. And this in turn implies that the "invention" of the eukaryotes was probably not a single event.

A Short Digression on Classification and Evolutionary History

The prokaryote-eukaryote dichotomy has a further lesson for us. For the have/have-not division we see here bears a message about fundamental principles of tracing genealogies—half of life's story. And in turn, it is upon the successful tracing of lineages, of defining and recognizing such pure "natural" groups, that we base our names of groups of living things—the naming of organisms; our species *Homo sapiens*, our Order Primates, our Class Mammalia, our Subphylum Vertebrata, our Phylum Chordata, our Subkingdom Deuterostomia, our Kingdom Animalia, our Superkingdom Eukaryota. Every single organism, past and present, belongs to a species; every species in turn is part of a genus (a series of closely related species), every genus is part of a family, every family part of an order, and so forth. We meet dogs at the level "Mammalia"; we share with insects the fact that we are multicellular, coelomate (possessing a body cavity) animals. We share with rose bushes the fact that we are eukaryotic; we share with bacteria the possession of DNA, RNA and a number of products elaborated by those molecules.

Sharing features in such a way provides a means of recognizing and defining such pure genealogical lineages. All vertebrates with hair, three middle ear bones, and mammary glands are mammals, for example. Darwin used this nested pattern of resemblance linking up all elements of life as a major demonstration of the fact of evolution: for if all organisms are descended from a single common ancestor, and if there has been divergence along the way, any new anatomical feature that evolution "invents" along the way will be passed on only to descendants of the original species in which the novelty first appeared. The same feature will not show up in other lineages that had already split off from the common ancestor. Thus we can search for these evolutionary novelties, and hypothesize that all organisms found to possess a certain feature (like hair) are related in an evolutionary sense—they are descended from a single common ancestor that lived sometime in the remote geologic past.

Tracing such groups gives us half of our story of the history of life. But on a more pragmatic level, we need to know what those groups are, and in particular we need to *name* them, simply to be able to talk about them. We need collective names simply to communicate about organisms; we speak of the "birds" in the trees, of the "elephants" at the zoo, and so forth. There are two living species of elephants, one of them (the Indian elephant, *Elephas maximus*) a close relative of the extinct mammoth of the Pleistocene (Ice Age), the other (the African elephant, *Loxodonta africana*) more closely related to extinct African ancestors. Both are joined, however, by being the only two living species of the Family Elephantidae, and "elephants" is merely a colloquial English word for a natural group of mammals—the Family Elephantidae. It would be intolerable if we had to enumerate each species (or each organism within a species!) when we wanted to convey something about a large array of organisms—all trilobites, say, or all mammals. Or even all elephants.

But it should never be forgotten that those collective names, especially in their formal, latinized zoological versions (*Aves* instead of "birds," for example) can, and usually do, represent natural branches of the genealogical tree of life. There is a real basis for those names—they are *not* arbitrary names of convenience. But then reflect on that prokaryote-eukaryote division once again. I called them the "have-nots" and the "haves." There is nothing wrong with "Eukaryota." It is based on a number of features (Margulis lists nine) common to the cells of all creatues included in the group—and found in no other organisms. The trouble begins when we turn to the bacteria—the "Prokaryota"—and see that they are defined strictly by the absence of those features we use to recognize the "Eukaryota." There are no *positive* features which join bacteria into a natural group—they are simply the primitive organisms that lack the advanced characteristics later added in the evolutionary history of life. And sure enough, recent research has revealed a host of different lineages within these primitive "bacteria," including a range of strange physiologies. Some are anaerobic, suited to the conditions of a primitive atmosphere lacking free oxygen (O_2)—a situation persisting until approximately 2 billion years ago and found now only in certain special environments. Others utilize sulfur instead of O_2. The prokaryotes are what taxonomists informally call a "garbage bag" group—everything left over after clear lineages are identified.

So, too, with the vertebrate-invertebrate dichotomy—discussed at the end of this chapter. Paleontology contains, in addition to "micropaleontology" (the study of microscopic fossils) and "paleobotany," the fields of "vertebrate paleontology" and "invertebrate paleontology." What are vertebrates? That's easy—all multicellular animals possessing a "backbone." What are invertebrates? That's not so easy—they are all those groups of multicellular animals that *lack* a backbone. Again, a group defined by the absence of features that characterize another, supposedly coordinate group. As you might expect, vertebrates are a natural, coherent evolutionary group whose genealogy can be studied successfully with an amalgam of information on both its fossil and modern representatives. We cannot say the same for the invertebrates, even though I freely admit membership in a department of that name at the American Museum of Natural History. Invertebrates are simply not a coherent, natural genealogic group—as we shall soon find out.

So names are important, but it is also important that the names we use be based on positive information, evolutionary novelties that serve to demarcate the evolutionarily natural groups. The entire history of systematics (the branch of biology and paleontology concerned with the recognition and naming of natural groups) can be read as a painstaking trudge toward the elimination of groups such as "Prokaryota" and "Invertebrata" in favor of groups in which we can have some confidence that their genealogies are "pure." There is still much work to be done along these lines, and some of the examples we will encounter as we look at life's history may well come as a surprise.

Evolutionary Pulses in a World of Precambrian Algae

But let us return to life's first 3 billion years. It was all "bacteria" for the initial 2 billion years; when "eukaryotes" appeared (to judge from apparent nuclei within some of these tiny microfossils), the basic look of the fossil record remained substantially the same—superficially barren, devoid of anything but tiny spherules and microscopic "cigars." Save one sort of feature— and that is very large indeed. These are "stromatolites," cabbage-shaped, wavy lumps that, on close examination, look like finely spaced laminae—thin plates interrupting the regular bedding of the sedimentary rocks in which they are embedded. The laminae, on still closer examination, are indeed particles of sediment arranged layer upon layer. Luckily, such structures are still being formed, as in the famous mounds in Shark's Bay off the northern coast of Australia. These mounds are formed of daily growth layers of a filamentous "blue-green alga," which is "prokaryotic" (true algae of various sorts being eukaryotic).

Thus not all of the Precambrian fossil record is ultramicroscopic. Large stromatolites are known as far back as 2.5 billion years ago, and they continue at odd intervals to appear in the record up to modern times. Yet they were never as abundant or diverse in overall shape as they were back in the Upper Precambrian, when they were the only large-scale form of life around.

It is in this early Precambrian world occupied solely by bacteria and algae that we find our first evidence of the episodic nature of evolutionary events. As I discussed in Chapter 1, evolutionary change is typically abrupt. Instead of the long periods of gradual modification we have become accustomed to expect in evolution, we usually find long periods of stability, interrupted at rare moments

FIGURE 3 Stromatolites, ancient cabbage-shaped mounds of laminated algal-trapped sediment, stand close together on the exposed surface of an ancient sea floor.

by the appearance of true evolutionary novelty. Surely the greatest event in the history of life was its very invention: there is a tremendous difference between a collection of organic molecules unable to abstract the energy needed to catalyze their own replication and an organized system that can do precisely that. It is easier to get from a bacterium to an elephant than from a soup of complex but unorganized molecules to a bacterium. It is very much as if life, once it got going, was content to leave it at that—to put it teleologically. Single-celled, microscopic bacteria simply exploited a number of basic metabolic pathways to obtain energy, and occasionally to divide. There was little distinction between maintenance of the body proper and reproduction—simple fission, division of the body. The DNA was not segregated from the rest of the cellular system—and reproduction and economics were closely interrelated. As life went on, this was to change—except, of course, in the "prokaryotes" still with us.

It is not entirely clear precisely when the complex symbiotic systems of the eukaryotic cell finally appeared. But why such an invention might have caught on, i.e., showed some survival value, is not hard to fathom. Eukaryotic cells show a remarkable division of labor, with organelles for ingesting substances, others for converting molecules into energy, still others acting as assembly plants for the building of proteins. The DNA is organized into coherent chromosomes and isolated from the rest of the cell, ensconced within a nucleus. Communication is via RNA, which copies the DNA and moves to the assembly sites (ribosomes). Reproduction—DNA making more of itself—is now isolated in the nucleus, more than ever set apart from the economic matter of protein synthesis.

We know all of this, of course, from studies of living organisms. But we learn of the pace of change in these systems only from the fossil record. And right away we see a few things that are utterly typical of all of life's evolution right up to the present moment. We see that inventions—of life, of the eukaryotic cell—apparently happen relatively abruptly. We must take care, since the fossil record is extremely scanty for these incredibly ancient, tiny fossils. And given the gappiness of the very formation of the fossil record, we must be prepared to admit that abrupt may still translate into millions of years. But abrupt it nonetheless seems—given the almost endless eons of time when things seem to remain pretty much the same. We will encounter this pattern again and again, though the tempo picks up as life becomes more complex. But here even the relatively skimpy, if tantalizing, early record seems to tell us that evolutionary events are few and far between, and the usual pattern is quiet, business-as-usual, interrupted only on odd occasions by the birth of something new.

And we see something else: that the invention of the new does not entail, as a matter of course (or even as the rule), the death of the old. There is no inexorable law of progress built into the history of life. Evolution is not a matter of constant improvement over old ways. True, living forms have become more complex over the past 3.5 billion years. That is, there are *some* organisms now alive that are more complex than any alive, say, a billion years ago. But extinction has claimed very few major groups of organisms—and apparently no truly major divisions of life. After all, we still have bacteria, as anyone who has had gastroenteritis can attest. In fact, it is very much still the Age of Bacteria, even though we tend to stick such labels on expanses of geologic time when nothing much but the organisms in question (bacteria in this case) were around.

What really happens is that alternative ways of dealing with both economics and reproduction are occasionally "invented" in one branch or another of living forms. Often, the novelty involves a division of labor, hence an apparent added complexity, as in the invention of multicellular animals and plants, in which cells perform one function, such as obtaining O_2 from the environment, while others are busy absorbing nutrients from the surrounding medium. No longer need one cell do all things. As long as environments persist in which a single cell equipped to perform all those fundamental tasks can exist, however, there is no reason for such a simpler system to disappear just because a more complex system comes along and takes up residence too. The ecological history of life is really concerned in large measure with the competition for resources—so sometimes one kind of organism will outcompete another, driving the other away. But on a gross scale, no entire array of more primitive forms has yet ceased to exist simply because a more complex configuration has arrived on the scene. That is what the fossil record says in general, and what the early and almost agonizingly slow first 3 billion years of life's history has to tell us.

The Next Step—The Sudden Appearance of Complex Animals

We tend to think of ourselves—*Homo sapiens*—as perched upon the very top of the evolutionary tree. Vain as it may be, there probably is a measure of justice to this conceit. But there are *two* main divisions of advanced animal life. The differences between the "protosomes" and "deuterostomes" are fundamental features that appear in the early stages of embryological development. For example, the opening ("blastopore") in the ball-like embryonic phase called the "blastula" is maintained throughout development and becomes the mouth of protostome animals. Hence their name: "protostome" means "first opening." In deuterostomes, a second opening is formed later in development to form the mouth, while the initial opening forms the anus. Along with echinoderms (starfish and their kin) and a number of fairly obscure worm-like groups, we are members of the Deuterostomia. By far the bigger group—in terms of sheer numbers of organisms as well as species—are the protostomes. Here we find segmented worms (earthworms, leeches, and the often beautiful polychaetes of the briny deep), mollusks (clams, snails, cephalopods, and kin), and the arthropods, which include the most numerous of all multicellular organisms—insects—plus crustaceans, spiders, and a host of less well known relatives. Though there are other protostome groups as well, annelid worms, mollusks, and arthropods are the most familiar. And they are every bit as anatomically complex, as highly organized, as are vertebrates. From such a vantage point, there really is no more reason to declare our own species the apex of evolution than, say, a katydid. Indeed, if sheer number of species is any indication of "success," we would have to admit that arthropods have it all over the vertebrates.

Trilobites are arthropods. In fact, they are the most primitive kind of arthropod known: most arthropods develop some degree of differentiation in the appendages ("legs") that project from their bodies. Crustaceans, for example, typically have legs divided into two branches; under the head, their legs have been modified into jaws plus associated manipulative organs involved in feeding. Farther back, the legs are variously used for walking, swimming, food gathering, and respiration. We know that trilobite legs, though extremely rare as fossils, were virtually identical from the mouth back through the tail region. Very primitive indeed—and just what we might expect, as trilobites are the very first arthropods to show up in the fossil record.

Yet on the other hand, trilobites are *not* what we would expect to find as the very earliest kinds of fossils: primitive as trilobites might be *vis à vis* other arthropod groups, the Phylum Arthropoda is as complex as any group of animals yet to have appeared in the history of life. But that is precisely what the fossil record seemed to show for well over a century of active paleontological investigation. For earth history really is divided sharply between a long period (nearly 4 billion years), in which stromatolites are the only readily visible traces of ancient life, and the period of the last 600 million years, with its rich and incredibly diverse, conspicuous array of life forms. Paleontologists have always wondered how that fauna got going. Concentrating their efforts down near the very earliest known fossils, they tried, largely in vain, to square what they saw with the expectations of evolution: we should find relatively simpler animals first; indeed, we should find a progression of forms through time, linking the simple with the more highly evolved. We shouldn't find trilobites at the very base. Yet until recently, trilobites were just about *all* that we found.

The Base of the Cambrian: Hunting for Trilobite Ancestors

The Romans called what is now Wales Cambria. Adam Sedgwick, Charles Darwin's tutor in geology, spent many years in Wales mapping the sequence of rocks and noting their mineralogic and fossil content. Around the Harlech dome, in the oldest sedimentary rocks of the region ("oldest" simply because they appeared to underlie all other sediments), Sedgwick collected trilobites. It was simply a matter of observation—a description of the way the world just is—that the earliest fossils, at least in Wales, are trilobites. Other geologists were finding trilobites too, and trilobites very similar to the Welsh species, in other localities in Europe and North America (and later in Asia and Africa as well). In the nineteenth century, Braintree, Massachusetts, produced a number of huge specimens of *Paradoxides*, very closely related to trilobites found in Newfoundland, Sweden, and Wales.

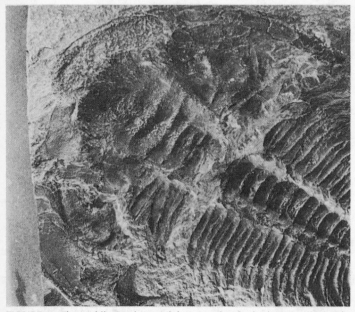

FIGURE 4 The Middle Cambrian trilobite *Paradoxides harlani*. A giant nearly two feet long when complete, this specimen was collected in Braintree, Massachusetts, in the first half of the nineteenth century.

As luck would have it, Sedgwick's Cambrian rocks were not complete: the very earliest segment of Cambrian time is not recorded by marine fossil-bearing sediments in that part of the world. But the story, in a sense, remained the same: even when we go lower—as in Morocco or southern California, well below the beds bearing *Paradoxides* and other trilobites typical of the Middle Cambrian—right down to the very base of fossiliferous rocks, we still find mostly trilobites. Now the earliest trilobites known are the olenellids; moreover, we are fairly sure that the earliest fossils from these far-flung regions are nearly of the same age because of the close similarity of specimens the world over. And there is even a measure of relief from the nature of these olenellids: they really are primitive for trilobites, particularly in the worm-like back end of their bodies.

Trilobites in general have heads (cephalon being the technical term) that are followed by a series of from two to over forty body segments hooked together as a flexible central body region (the "thorax"). A solid tail piece (or pygidium) brings up the rear. Olenellids have a tiny nubbin instead of a well-developed pygidium, and the thorax simply tapers down, in worm-like fashion, toward the rear.

We have been, metaphorically, walking down through time, through fossiliferous trilobite-bearing strata, down to the *Olenellus* or *Fallotaspis* zone—then nothing. Peering over the edge of the abyss of time, we have been seeing nothing back beyond the earliest trilobites, save stromatolites and microscopic bacteria and algae. But recently the picture has changed. There is more, and some of the missing pages in the book of life have been filled in—and the emerging story is not exactly what we might have expected.

A Window on Latest Precambrian Life—An Ambiguous Link to the Cambrian Fauna

In the late 1940s, again in Australia, where some of the earliest rocks and fossils occur, an unusual collection of fossils came to light. In the reddish, quartzy beds below the Cambrian rocks of the Ediacara Hills in the Flinders Range, a variety of impressions, obviously of organic origin, began to attract attention. The fossils are somewhere in the neighborhood of 670 million years old.

Immediately we see that there is something strange about this "Ediacaran" fauna. We might expect, for example, that any newly discovered collection of fossils below the Cambrian would contain the forerunners of the trilobites, brachiopods, and echinoderms that we see in the earliest full-blown Cambrian assemblages. Instead, the Ediacaran fauna contains a mixture of definitely recognizable sorts of creatures—mainly worms, but also various relatives of "soft" corals (known in modern seas informally as "sea pens")—plus a variety of more puzzling animals difficult to place in the spectrum of known forms of life. One thing stands out: ecologically the fauna is unlike anything we see in Cambrian rocks. Indeed, it is unlike nearly all subsequent fossil assemblages. You have to go to modern shores to find comparable collections of soft-bodied creatures.

And that is perhaps the one most important aspect of the Ediacaran assemblage: all of the fossils in it are impressions of animals that lacked any hard parts (although some paleontologists have claimed to have detected traces of calcium carbonate, the principal ingredient of most shells, on *Tribrachidium*, which closely

FIGURE 5 The Ediacaran fossil *Dickinsonia*, perhaps a forerunner of the segmented worms.

resembles primitive echinoderms). The reason why there is a rich record teeming with fossils from the base of the Cambrian onward is simply that organisms with hardened organs—*skeletons*—had appeared. Now that finally an earlier fauna has turned up, one would predict that the creatures in it would for the most part be soft-bodied. Prediction confirmed. But one would also expect these soft-bodied organisms to be the recognizable forerunners of the hard-shelled trilobites, mollusks, and echinoderms of the Lower Cambrian. Wrong on that one: we get mostly worms, jellyfish of various sorts, those sea-pens, and a variety of creatures even more difficult to identify.

Disappointment, if not consternation—but only in conventional evolutionary terms. The point remains: There were no hard-shelled invertebrates prior to the very time that we first pick them up in the record. Yet multicellular organisms *had* evolved, as suspected, prior to the earliest trilobites. Soft-bodied creatures are rarely preserved: the Ediacara fauna is actually the first (discounting some remarkable occurrences of the Precambrian microflora) "window" onto a world that is *usually not preserved*. We are lucky to see it at all. And if we are sampling soft-bodied invertebrates, why not expect them to be the sorts of animals we know to be soft-bodied today? For that in the main is what the Ediacara fauna is telling us: there were multicellular animals in the Upper Precambrian; they were soft-bodied, and they resembled in general the sorts of soft-bodied organisms we see today.

Nor is the Ediacaran fauna utterly unique. Other occurrences of soft-bodied forms—some as long as six feet—are known from Newfoundland, England (*Charnia* is a "sea-pen" from the Charnwood forest), and southern Africa. None of these occurrences is anywhere near as prodigious in sheer numbers of fossils or kinds of organisms present as we find in the Ediacaran assemblage. Yet as brief glimpses into life as it was 670,000,000 years ago, they confirm the impressions of multicellular animal life we gained from the Ediacaran fauna.

Some paleontologists, most notably Adolf Seilacher of Tubingen, West Germany, have recently re-evaluated the Ediacaran fauna, stressing its "strangeness," the fact that few, if any, of the fossils are easily seen as belonging to familiar invertebrate groups. Seilacher brings out an important point: the very nature of the pre-trilobite, Upper Precambrian fauna is ecologically and genealogically quite distinct from any fauna that follows it. Yet we must remember that all of life is interrelated—which, as Darwin saw, translates into the very pragmatic realization that absolutely every kind of organism, whether still with us or extinct, is related to every other in a complex, yet unambiguous, fashion. It is easy to assert that the organisms we see in the Ediacaran-style faunas are not close relatives of any kinds of organisms that we know, but surely probabilities lie on the side of genealogical interconnectedness: it is a simpler and hence more reasonable position to see those soft-bodied animals as relatives of soft-bodied forms of more recent times to which they show definite patterns of similarity.

The confusion over the Ediacaran fossils comes, I think, from the indisputable fact that only a few bear any resemblance at all to the sorts of hard-bodied fossils so common once we reach the base of the Cambrian. *Spriggina* is a worm-like fossil with a fairly well differentiated head region—and thus a possible early forerunner for trilobites. And I have already mentioned *Tribrachidium*, a flattened disk with three spiral arms and, some say, an occasional trace of calcium carbonate over its surface. Modern echinoderms all have "pentameral symmetry"—think of starfish, for example, with their five legs (or, more rarely, multiples of five legs). Under each starfish arm lies a groove (ambulacrum) lined with tube feet which are used for grasping both prey and the seafloor. The tube feet are operated by an internal circulatory system—the water vascular system—that regulates the internal pressure of the water filling each tube foot.

FIGURE 6 *Tribrachidium*, a possible primitive echinoderm from the Ediacaran fauna.

FIGURE 7 *Spriggina*, an Ediacaran fossil that appears to possess a head and a segmented body—and thus possibly was allied with the segmented worms or early forerunners of the arthropods.

Echinoderms of the Lower Cambrian, such as the helicoplacoids found in the White-Inyo Range of the California-Nevada border, have a single ambulacrum. Studies of the nervous system of both modern and fossil echinoderms have recently shown that the primitive symmetry is probably three-fold rather than five-fold, which was a later development. *Tribrachidium*, remember, has three arms. In fact, *Tribrachidium* looks very much like an edrioasteroid—one of a group of disk-like Early Paleozoic echinoderms that begin life with three, but soon add two to give the full, expected echinoderm complement of five ambulacra, arranged (as in *Tribrachidium*) in impressive spirals across the flat surface of the body.

But that's it. Otherwise the Ediacaran fauna contains not a hint of the hard-shelled creatures yet to come. Hence Seilacher's insistence that we are dealing with what could really be considered a separate radiation, an "evolutionary fauna" of earlier vintage quite unrelated to the appearance of complex life at the base of the Cambrian. But maybe we were simply wrong in supposing that, if and when we found Upper Precambrian multicellular animal fossils, they would do the expected thing and look like primitive forerunners of the Lower Cambrian trilobites, echinoderms, brachiopods, and mollusks. The search for those "missing links" in that long period which the evolutionary development of all those complex animals seemed necessarily to demand, is based on some logical deductions that arise from the belief that evolution always takes a very long time and proceeds in a linear path from simple to ever more complex.

Rethinking the Origin of the Cambrian Fauna

Unfortunately, the premise of slow, linear evolution does not apply, and the prediction that any soft-bodied Precambrian fossil ought to look like a prototype of later, hard-shelled fossils was wrong. Nor is this simply the benefit of 20/20 hindsight. From the earliest discussions of the advent of the Cambrian fauna, there have always been a few paleontologists pointing out what strikes me as an obvious fact, albeit one with some confusing "Catch-22" implications: you simply cannot have a brachiopod, say, for example, or a trilobite or an echinoderm, *without* those hard skeletal parts. A brachiopod shell consists of two valves interconnected by a series of muscles that open and close the shell. The "soft parts"—the feathery lophophore, for example, which filters food particles from the surrounding water—are all attached to the shell, withdrawn inside when the shell is closed, and together form a portion of the organism that could by no means be expected to function normally without the shell—let alone survive predation. In other words, a "thought experiment" in which the hard skeletal parts are subtracted from each of these common sorts of Lower Cambrian fossils produces, in every case, an anatomy with truly bizarre lumps of soft tissue quite unlike any known functional soft-bodied organism. In nature, by contrast, the soft anatomy is arranged in conjunction with the skeleton. Without a shell you simply cannot *be* a trilobite or a brachiopod. (Aha, I hear the cry—you ignore the fact that some snails are shell-less; yes, I say, that is true, but these are secondary losses reflecting the invention of still newer ways of making a living as a snail, without the aid of a hard skeleton. An escargot plucked from its shell looks rather different from its close slug relatives!)

There is a way out of this paradox—for that's what it is: if we took the argument too far, we would have to conclude that trilobites never had soft-bodied ancestors. A complex series of muscles can push and pull against a relatively thin, poorly mineralized skeleton *if* the entire organism is relatively small. This possibility fits the situation beautifully, for small, poorly mineralized shells do not lend themselves readily to making fossils. The scenario is plausible, but if anything a bit too convenient as an explanation for the lack of a smoothly progressive series of fossils leading up to the trilobites and their cohorts that show up at the base of the Cambrian.

Appealing as the small-is-primitive explanation may be, the explosion of life at the base of the Cambrian is best understood in a dual theory that speaks of genealogy—the derivation of new sorts of organisms from old—but also sees the "event" as ecological: after all, a wide variety of different sorts of organisms all show up with hard skeletons at roughly the same time: maybe there was something about the Earth's overall physical environment that enabled an assortment of groups, small or not, to "invent" hard skeletons: if so, perhaps we can see that some kinds of organisms merely added a hard outer covering while other groups' acquisition of a hard outer covering provided the opportunity—even the necessity—to radically reorganize their soft parts, leading to the emergence of true novelty in the animal kingdom.

Apparently the Earth's atmosphere essentially lacked oxygen until about 2 billion years ago—the age of the earliest banded iron-rich sediments, in which the iron is oxidized into both ferrous and ferric states. Some twenty years ago, C. L. Berkner and L. C. Marshall proposed a link between the rise in O_2 content of the Earth's atmosphere and the development of complex forms of life. They hypothesized that, despite the switch to an oxidizing atmosphere, the amount of oxygen in the atmosphere, hence the amount dissolved in sea water, was far too low to support complex life. Initially, oxygen accumulated through simple outgassing from volcanic vents; later, as photosynthesis by single-celled organisms went on, life itself became the dominant source of free oxygen. Note, too, that without oxygen (O_2) there will be no ozone (O_3), an important ingredient of the upper atmosphere, as it shields the Earth's surface from ultraviolet radiation harmful to life. Berkner and Marshall suggested that near the end of the Precambrian the O_2 level rose to a critical point (though still only 1 percent of the present level)—to a level at which complex organisms could flourish.

Similarly, paleontologist S. M. Stanley has suggested that the monotonously simple ecosystems of the Precambrian might be expected to display rather low diversity (meaning numbers of different species). Each species, after all, will be doing roughly the same thing—photosynthesizing. It stands to reason that only a few different species will tend to dominate, crowding out most potential competitors. Only when some simple animal-like forms—single-celled forms that ingest other organisms, such as amoeba-like protozoans—show up and ecological life becomes more complex, are there (by definition) more niches available to fill.

When we take the notion of a chemical "threshold effect" permitting more complex life, together with Stanley's suggestion that active browsing on those algal mats that dominated so much of the Precambrian seafloor would create a more heterogeneous bottom to house a greater variety of organisms, we have some signs pointing the way to understanding that great Lower Cambrian proliferation of organisms. It is very much as if, once the opportunity arose and life finally had started along the road to greater complexity, the biologic world rushed in to fill a vacuum. Some creatures really do seem simply to have hardened their outer surfaces, leaving their soft inner parts pretty much as they were: corals (which, interestingly, don't appear until the Middle Cambrian, so far as is known, and in any case do not really become abundant until Ordovician times) seem to have simply added a calcium carbonate outer layer to the simple structure of a polyp (the soft cylindrical body, as seen today in sea anemones; we will see more of coralline anatomy later as the corals become more of a force in life's history). Trilobites, brachiopods, mollusks, and echinoderms, on the other hand, seem to represent radical reformations of ancestral anatomies: hard coverings are added, but immediately we find the internal anatomies reorganized. Trilobites were undoubtedly derived from some annelid worm-like forerunners; as arthropods, they devloped jointed legs and reduced the internal segmentation of the body as the major departure from segmented worm anatomical organization. Mollusks, too, are thought to derive from the annelid worm stock, losing virtually all traces of internal segmentation; in contrast to arthropods with their jointed outer skeletons, the mollusk shell is an ever-growing, solid one- or two-piece structure (with the eight-plated chitons the sole exception). No one knows where the echinoderms came from—their greatest similarities are with the larvae of some obscure groups of marine "worms." Of all complex animals, echinoderms are definitely the strangest we have on Earth, as we shall see later in this chapter.

Thus we have abandoned a gradual model for life's grand diversification in favor of a picture of rapid deployment. And so it is a bit ironic to discover that, in the past twenty years, paleontologists have found a whole series of hard-shelled invertebrates—mostly mollusk-like creatures— *below* the earliest known trilobites!

Filling in the Gap: Events Just Below the Cambrian

Siberia boasts more than salt mines and Gulags. There is also a thick sequence of sediments, the so-called Vendian, leading up to the base of the Cambrian, and it is from there that the best array of true hard-shelled fossils from the Upper Precambrian has been assembled. It is now apparent that the hard-shelled invertebrate fauna really did take some 10 or 15 million years to become fully established. But it is all a matter of scale: 15 million years is a brief interval compared to the eons of bacterial history that came before, and the hundreds of millions of years that complex life forms have inhabited our planet since their advent. We still do not see the gradual development of the by-now familiar Lower Cambrian groups; the sudden proliferation of trilobites, echinoderms, brachiopods and the like continues to confront us. But ecologically, we do see that the entire Lower Cambrian "ecosystem" was assembled piecemeal over a considerable span of time, rather than overnight in a sudden creative spurt "full blown from the brow of Zeus."

But one needn't travel all the way to Siberia to experience this more staggered sort of pattern of the advent of complex life. Many of the ranges that line the Great Basin of North America (which is largely confined to Utah and Nevada) contain magnificent exposures of Cambrian rocks. In the extreme southwest portion of the Basin, from the California White-Inyo range eastward into Nevada, the lowermost Cambrian is well developed. Though this area has long been known as a prime source of olenellid trilobites, more recent work has revealed a thick sequence of uppermost Precambrian sediments leading up to the basal trilobite zones. Stromatolites abound in these sediments, but so do some oddities, such as the tiny shell (mollusk?) known as *Wyattia*, a rather simple little curved tube (named for retired Berkeley paleontologist J. Wyatt Durham—an example of how some of those very formal-seeming "Latin" names come to be). In one narrow draw leading up through a veritable wall of sandstone, exposing hundreds of feet of Lower Cambrian rock in the Inyos, relatively simple trails left by some unknown multicellular creatures (worms?) appear near the bottom. As you climb, the so-called trace fossils seem to become more complicated, perhaps reflecting the advent of more sophisticated modes of behavior. The first trilobites put in their appearance hundreds of feet farther along the trail, higher up the pile of sedimentary rock, and thus probably some millions of years later. Trace fossils—footprints, tracks, trails, and the like—occur throughout the fossil record, but for reasons that are not always entirely clear, rarely occur in the same beds with actual body fossils. However, some of these trace fossils are identical with the traces that trilobites are known to have left as they scuttled across the seafloor. Thus we cannot be witnessing the transition from some unknown sort of soft-bodied organism to trilobites in this Inyo sequence. But the sequence does serve to remind us that the fossil record is *not* complete: the trace fossils indicate that there were trilobites living in that patch of real estate before the ones that left their skeletons for us to find as fossils.

I have pursued the advent of the Cambrian fauna at some length because it is the third great event (life's invention and the appearance of eukaryotes being the first two) in life's history and an outstanding and thought-provoking area of paleontological research and theory. The actual pattern of this event requires careful analysis: a mere cracking of more rocks alone won't tell us what happened. And that is how it should be: real science is a complex combination of data—observations—and ideas that seem to explain those observations. But there is a thin line between data and theory; ideas become data when they are repeatedly confirmed (e.g., "the earth is a sphere"). And we need these observations to keep our ideas about the world "honest." Yet without ideas, who would bother climbing all those cliffs, trying to document in locality after locality what really happened in those 20-odd million years when life as we know it really got underway? Only with some notion firmly in mind—that life has evolved, that complex life seems to have arisen relatively suddenly after an incredibly long period when life remained very, very simple—does it make any sense to climb these rocks and collect all these fossils.

Biological nature really does seem to abhor a vacuum. Once hard-shelled multicellular organisms appeared on the scene, they rapidly deployed into a vast array of anatomical shapes and sizes, and therefore presumably into a rather heterogeneous array of ecological niches. Yet it is not quite true that the evolution of complex life was over by the end of Lower Cambrian times. As J. John Sepkoski, a paleobiologist at the University of Chicago has pointed out, the Cambrian invertebrate world had a somewaht different complexion from that of the world familiar to paleontologists who deal with the last 500 million years or so of geologic time. It is as if complex life had a shake-down cruise in the Cambrian; after extinction had weeded out many of the Cambrian actors, the drama was renewed in the ensuing Ordovician Period with a somewhat altered cast. As we shall see repeatedly throughout life's long history, it has been extinction that has opened many of the opportunities for evolutionary innovation. Trilobites, kings of the Cambrian seas, were never to regain their dominance; it is the vicissitudes of trilobite history that contribute most heavily to the contrast between Cambrian and later life in the sea.

An Expedition in Search of the Cambrian World

Antimony Canyon takes its name from a small but workable deposit of ore-bearing minerals that proved a valuable source of that rather rare metal during World War II. The canyon is really a shallow gash, one of many cut into the west face of the Wellsville Mountains north of Brigham City in northern Utah. The narrow jeep trail that takes you up the precipitous slope looks out over magnificent Great Salt Lake. The Wellsvilles themselves are remarkable: the northerly extension of the Wasatch range (the most westward range of the Rocky Mountains in the continental United States), the Wellsvilles are known as the Earth's *steepest* range: their ratio of height to width measured through their base is greater than that recorded for any other mountain system.

The Wellsvilles are a good place to get acquainted with the Cambrian world. Beyond that massive saline lake—ironically a mere remnant of the vastly greater Lake Bonneville of the Ice Ages—begin the mountains of the Great Basin and Range system. Stretching across western Utah and all of Nevada to the Sierra Nevadas, most of the mountains in this desert run roughly north-south. They are there because in a phase of crustal stretching that became intense only some 3 or 4 million years ago, what are now the valleys dropped down as huge crustal blocks, leaving what are now the mountain ranges (with such colorful names as the "Wah-wahs," "Crickets," and "Confusion Range") as positive blocks jutting through. The rocks within these ranges are a varied lot. Many are Paleozoic sedimentary rocks, and the Cambrian is exceptionally well represented—particularly in southwestern Utah, where the House Range produces the little black trilobite *Elrathia kingii* in such profusion that it is sold, generally at modest prices, in rock and curio shops the world over. For Cambrian trilobites, especially Middle Cambrian trilobites, Utah just cannot be beaten.

But we weren't climbing the Wellsvilles just for the view. The Cambrian occurs in the Wasatch range, too, and Lloyd Gunther, one of a relatively small but important group of dedicated amateur paleontologists in the United States, was leading me and two colleagues from Utah State University in search of the Spence Shale, with its variegated fauna of trilobites plus the elusive, rare "eocrinoid" echinoderm *Gogia*. Gunther was one of the few who had had the patience to seek out the Spence up that canyon (and all the others in the region); it was he who had discovered the rich caches of *Gogia* and its intriguing trilobite companions. Indeed, it was Gunther who discovered the very *existence* of eocrinoids in the first place. Parking the jeep near the entrance of the (now abandoned) antimony mine, we shot up without further ado for an additional 1,000 feet or so—with Gunther (who is retired) barely puffing, while I, a city-slicker forty-ish paleontologist accustomed to long months in the laboratory, frankly had a tough time (I stopped often to admire the many Sego Lilies, Utah's state flower, that dotted the hillside).

Sedimentary rocks exposed in mountainsides are often far from regular, flat-lying, undisturbed sequences. On Mount Everest, for example, rocks of Upper Permian age (the end of the Paleozoic) sit *on top of* younger, Cretaceous sediments. The reason: true mountain ranges, such as the Rockies and Himalayas, represent a crumpling—thus a narrowing—of the Earth's crust. It is like taking a small throw rug and tossing it on the floor before you: invariably it will end up in a series of accordion-like pleats, much shorter overall than the rug's "normal" length when stretched out flat. In mountain building, as the rocks are pushed laterally together, sometimes those huge crustal pleats crack open. One side will then ride over the other, to be discovered millions of years later as an anomaly in the usual sequence of rocks and fossils regularly encountered in the undisturbed sequences that blanket the stable interior regions of continents. And that, confusingly enough, was the way the bluish-gray Spence Shale appeared amidst a huge pile of rather unfossiliferous rock. It was particularly strange to sit there splitting rock in the quarry, gazing across the narrow chasm of Antimony Canyon, and not see the same Spence shale layer at the same level on the other side, the way you most definitely see the same layers exposed on both sides of the Grand Canyon.

Trilobites are nearly everyone's favorite fossils—they have heads, usually with eyes, and bodies; they really *look* like an animal, more so than most fossils, even dinosaurs (which usually show up as isolated bony fragments rather than entire skeletons). And I am a trilobite paleontologist. So naturally, after the trouble of actually getting up to that little Spence Shale exposure on the southern face of Antimony Canyon, I was after trilobites. Some rather large specimens of various species (*Athabaskia wasatchensis*, for example) make especially good trophies, though much is to be learned from various isolated bits and pieces of trilobites—in fact, every specimen is potentially worth collecting and adding to the collections back at the Museum. But as luck would have it (and fossil collecting is every bit as much a matter of luck as of skill), that beautiful July day was to be echinoderm day for me: I collected about ten of the rare eocrinoids for the American Museum; I added *no* trilobites worthy of particular notice. But the echinoderms were also exciting, and also very good for the paleontological soul, for in painting a picture of Cambrian life it is tempting (and traditional) to speak endlessly about trilobites. They were the dominant form of life in those ancient seas; though they remained an important element of life in the seas for millions of years thereafter, they never were as abundant or diverse—or certainly as dominant over other life forms—as they

FIGURE 8 The trilobite *Athabaskia wasatchensis* from the Middle Cambrian Spence Shale, Antimony Canyon, Wellsville Mountains, Utah. Collected by Mr. Lloyd Gunther.

FIGURE 9 The eocrinoid echinoderm *Gogia* from the Middle Cambrian Spence Shale.

were for those first 75 million years or so of the history of complex life. But to focus exclusively on trilobites is to miss some of the more fascinating and, in a way, arresting aspects of the fiber of Cambrian life. For Sepkoski, I think, is right: life in the Cambrian seas is a bit out of kilter (not his choice of words) when compared with later marine life—which from the Ordovician onwards is very much like marine life as we know it today. And when I think of the odd-ball creatures that populated life's first attempt at complexity, I think first of the echinoderms. Echinoderms themselves are odd as animals go; Cambrian fossils make the echinoderm story even more strange.

The World of Echinoderms

Echinoderms hardly seem like animals at all. For starters, none of them has a head. I once watched a movie, part of a presidential address to the Paleontological Society by Porter M. Kier, until recently the Director of the Smithsonian Institution's Museum of Natural History. The movie, filmed in infrared with time-lapse photography, showed a technician turn off the normal overhead lighting in a room, then shut the door. Three regular echinoids (spiny, globular, "sea urchins") glided down the sides of an aquarium tank, converging on the middle. There, astonishingly, they jointly exhumed a flat sand dollar (an "irregular" echinoid—an evolutionary derivative of the more primitive "regular" echinoids). The three of them then proceeded, in a rather calm and matter-of-fact manner (if I may be permitted such anthropomorphisms), to munch their hapless kin out of existence. By "daylight," all three urchins were innocently back up on the sides of the tank. I felt as if I had witnessed an ecological event on Mars.

FIGURE 10 An irregular echinoid from the Eocene.

There are five major sorts of echinoderms (the name means "spiny skin") living today: starfish (asteroids), brittle stars (ophiuroids), sea urchins (echinoids), sea lilies (crinoids), and sea cucumbers (holothuroids). Starfish and brittle stars—equipped as both groups are with five "arms" (or multiples thereof)—are rather similar, bottom-dwelling carnivores. They hardly resemble the armless sea urchins; and the sea lilies look, as their name implies, more like plants than animals. Crinoids (sea lilies) become the dominant echinoderms of the Middle and Upper Paleozoic; their long stalks, composed of hundreds of disk-shaped columnals, are so common that they may be the dominant constituent of some limestones. Atop the stalk sits a cup (the calyx) housing the digestive and reproductive systems. From the cup emerge five feathery arms with cilia used to strain food particles from passing water currents.

FIGURE 11 *Devonaster*, a Devonian starfish.

The key ingredient in all this is the number five. Even the armless echinoids have five bands (ambulacra), where tube feet emerge from the shell to function, once again, as means of both locomotion and feeding. The water-vascular system that operates the cilia and tube feet of these otherwise dissimilar beasts is absolutely unique in the animal kingdom—another feature that unites the echinoderms into a natural evolutionary group. But pentameral symmetry, as we have already seen, was a relatively late invention in echinoderm evolution. Indeed, none of the five living classes of echinoderms goes back further in the fossil record than the Ordovician Period.

FIGURE 12 An Ordovician crinoid.

Even though modern echinoderms did not appear until the next geological period *after* the Cambrian, they are still relevant to our understanding of what was going on with these strange creatures in those earliest faunas of complex animals. If to the five major echinoderm groups ("classes" in the Linnaean system of classification), we add all the other sorts of extinct echinoderms—mostly known from the Cambrian and early Ordovician—we come up with a total of over twenty classes of echinoderms. In contrast, there are only six living and perhaps three or four extinct classes of mollusks, some seven or eight classes of Arthropoda (which contains by far the greatest number of animals of any phylum on Earth today), and the five traditional (and as we shall later see, outmoded) classes of the Subphylum Vertebrata (of our own Phylum Chordata).

So echinoderms tend to come in a relatively great number of classes. This is really only to say that there are few anatomical clues in the adult bodies of these creatures that point to other groups in the Phylum Echinodermata which might be their nearest relatives. Beyond sharing a few general features with all echinoderms (another important one being the microscopic structure of their skeletal plates), members of these different extinct classes are as different from one another as modern adult echinoderms in each of the extant classes are one from another. It is a real handicap, one that must be taken into account, that the fossil record does not for the most part allow us to study the course of development from fertilized egg to adult of these long-dead animals and plants. However, echinoderm plates do have growth rings, and by studying smaller fossils—animals that died while relatively young—it is possible to document some aspects of their skeletal growth during development. As we shall see, trilobites periodically shed their skeletons during growth, while mollusks live in ever-growing shells, carrying the history of their shell growth around with them throughout their lives. And sometimes, it is even possible to glean some insight into the embryology of the soft anatomical parts of these animals.

Most high-school biology students still learn the old phrase of Ernst Haeckel: "Ontogeny recapitulates phylogeny." It means that in the course of development, an organism will in some fashion "go through" various stages in the evolutionary history of its ancestors. When interpreted too strongly, for example, when we suppose that some intermediate embryonic stage represents some form of ancestral adult, we go too far: gill slits along the "throat" of a human embryo do not in any detail resemble the gill slits of any fish; human embryos, instead, resemble the embryos of other creatures, particularly other mammals, but also birds and other vertebrates. But when we say that "vertebrates are united by the possession of pharyngeal gill slits," it is our knowledge of embryology that tells us this; land-living vertebrates do not have gill slits in any postembryonic phase of life.

Thus, embryology heightens our awareness of the true distribution of anatomical features—items that give us the "evolutionary signals" enabling us to tell who is related to whom. Echinoderms simply remain an enigma with their apparent "higher-level" diversity that seems so out of phase with the patterns of evolutionary relationships typical of the other major groups of animals. But it has been suggested that in those early days, when life was, in a very real sense, just feeling its way, the echinoderms engaged in a truly explosive episode of evolutionary diversification, "experimenting" with a host of alternative body plans, variant versions of echinoderm anatomical organization. Some echinoderms were rooted, some cemented to the shells of others. And some crept around the seafloor. After all, many of life's more inventive moments seem to have come after extensive extinction events greatly depleted the stock of living things—literally opening the door of opportunity for the invention of new ways to solve old ecological problems. Perhaps it is not too surprising that at the very inception of complex life, we would get the oddest array of creatures ever seen. In fact, they only seem odd to us because they do not fit in as well with modern life as do their descendant denizens of the Ordovician and later seas.

And yet there's another odd thing about echinoderms. Though they are rare as fossils (except for the ubiquitous crinoid columnals of the Mid- and Upper-Paleozoic), when you *do* find a Devonian starfish, or a Cambrian eocrinoid—just about any fossil echinoderm imaginable—you very often find them in droves. I found about ten eocrinoids that day working the Spence Shale in the Wellsvilles. My companions found still others. And yet *Gogia* is known from but a few localities in the Middle Cambrian of Utah and Idaho. They are rare as hen's teeth (which are now known embryologically!), yet, when you *do* get them, they tend to be abundant. Just another sign of the oddity of echinoderm paleobiology, their *otherness*: though, interestingly enough, their patchy distribution in the fossil record does agree well with observations of modern echinoderms: starfish and ophiuroids, for example, very often litter the seafloor in dense concentrations.

A Thumbnail Sketch of Cambrian Life

There were other Cambrian organisms, of course. The earliest "reefs"—organically produced mounds on the seafloor—were created by archeocyathids. Long thought to be sponge relatives (many sponges have hard, calcareous skeletons; they are not all bath "spongy" in texture), archeocyathids now appear more closely related to certain forms of algae (which can also have hard calcareous skeletons). These organisms, consisting of one porous cone stacked within another, sat on the seafloor (either photosynthesizing or filtering food from the water—it is frustrating to be *so* ignorant of life's remote past). Like corals and some sponges, archeocyathids were often attached to one another, forming colonies that provided the basic framework for a hard "reef." And hard they are: I climbed my first archeocyathid reef in Lower Cambrian rocks well down near the base of fossilized life in Nevada.

Eyeing the hatchling Prairie rattlers which call that old reef home these days, I found it fascinating that Nature was building reefs so soon after the advent of complex life—and using organisms un-related to corals and sponges at that. And reefs are important. The rattlers today merely carry on a tradition: all manner of plants and animals find food and shelter in the nooks and crannies of reefs, and their presence draws the large predators who live around the reef face. Ask any scuba diver where he or she prefers to go to see modern marine life.

And not all of Cambrian life was so very strange: there were mollusks, but not in the number or diversity in which we see them today, or that brought them into prominence in Ordovician times. There were brachiopods, too, but mostly inarticulates lacking teeth and sockets along their hinge lines—a group still alive today, though numerically far less abundant than their cousins the articulate brachiopods, which proliferated again in the Ordovician to become the dominant form of Paleozoic marine life. There are rather small, nondescript orthid articulate brachiopods (often called "lamp shells" because of the resemblance some of them bear to the oil lamps of ancient times), starting in the Lowest Cambrian. But they are truly rare. Even the trilobites that so dominate the Cambrian world hardly match up with those that lived on the other side of the Cambrian boundary; though trilobites are still fairly common in the Ordovician, few post-Cambrian trilobite groups seem similar to those of the Cambrian; and sheer trilobite diversity and numbers of individuals dropped drastically after the Cambrian. Trilobites have much to tell us about the Cambrian world—both physically and biologically. For the advent of a good fossil record has profound, and very practical, geological implications: we can make much more sense of our rocks, sort them out, and tell what was going on at about the same time all over the Earth with far greater accuracy than we can in the nearly barren rocks of the Precambrian. But in taking the trilobites as a coherent group, we must follow them through to the end, which will also be our first foray above and beyond the Cambrian. We will glimpse sea life through what so far has been the greatest single span of coherent history: the post-Cambrian Paleozoic.

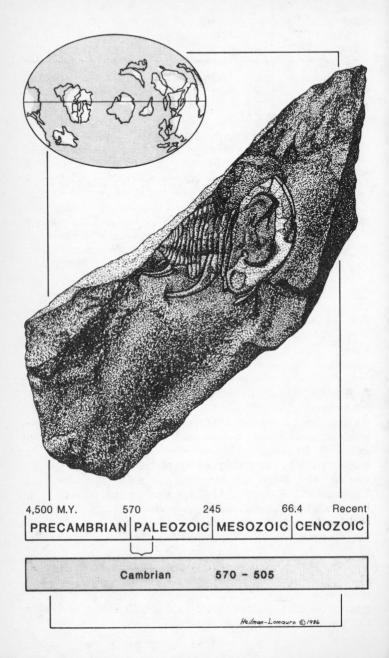

4,500 M.Y.   570   245   66.4   Recent

PRECAMBRIAN | PALEOZOIC | MESOZOIC | CENOZOIC

Cambrian   570 – 505

Heilman-Lomauro © 1986

3

The Trilobites: Cambrian Excursions and Later Diversions

Much of paleontological research, if not exactly of the armchair variety, takes place in the laboratory rather than in the more glamorous setting of the field. My own earliest experiences with fossils were almost embarrassingly remote from the personal collecting experience: snails from Oklahoma and Texas formed the basis for my first professional publication. Another early venture involved some rather unprepossessing little trilobites from Newfoundland. And though I had never been to any of those places, I soon found that I could range far and wide, over vast distances and back hundreds of millions of years—just by sitting patiently back in the lab digging specimens out of their rocky matrices and consulting what others before me had written about them and similar beasts from the past. This kind of research is not without its rewards. Occasionally you find out something about the world that would never occur to you on some remote outcrop, so different today, exposed as it is to the air, from what it was when it was plunged below the waves all those millions of years ago.

Cambrian Trilobites and the Evidence of Continental Drift

Those Cambrian fossils from Newfoundland were my first serious exposure to trilobites. I had only recently decided to become a trilobite specialist, and had the decidedly mixed blessing of *not* being apprenticed to a recognized expert in trilobite paleontology. I was pretty much on my own. And no sooner had I made that decision when a former professor, Marhsall Kay, a senior member of the Columbia geology faculty, asked me to identify some specimens he had dragged back from central Newfoundland over the past summer.

Kay was an interesting man. Born into a family in which geology was a tradition, he developed a theory of the structure and essential "behavior" of the Earth in the 1940s—a notion that made the connection between modern-day "island arcs" (such as Japan, the Aleutians, the Philippines, the outer archipelago of Indonesia) and the fully consolidated linear mountain belts that we see on the continents. Kay was a stratigrapher—an expert in the unraveling of sequences of bedded rocks and thus in the unraveling of the history of the Earth's crust. (It often galls the more biologically inclined paleontologists to hear the old saw "paleontology is the handmaiden of stratigraphy"—as if our sole *raison d'être* is to help stratigraphers correlate rocks by identifying their fossils. Yet it is nonetheless true that precise identification of fossils really *does* aid immeasurably in deciphering Earth history.)

Kay went into the most difficult, jumbled terrains of rocks and proceeded to make sense out of them. He especially liked the roots of old mountain systems—and Nevada and Newfoundland are especially good for perplexing melanges (in both the conventional and technical sense, a geological "melange" is a confusing welter of mixed rock types). Newfoundland is a geologically arresting place: on either side of this island Cambrian rocks are especially well developed. Moreover, they are laid out in fairly undisturbed fashion. From the western side, there are archeocyathid reefs and exquisite olenellid-bearing beds. From the eastern side, the Avalon Peninsula exposes a thick section of Middle and Upper Cambrian rocks, jammed with some magnificent trilobites.

For decades it had been known that the Cambrian trilobites on the two sides of Newfoundland simply did not match up. How could trilobites living a scant 300 miles apart possibly remain so different for so many millions of years? The problem was not age: many of those Cambrian rocks overlap in age. Nor could the explanation be purely "environmental"; it would be unlikely on the face of it that major habitat differences, supporting completely different trilobite faunas, would persist for so long. Moreover, the trilobites from western Newfoundland resembled others ranging clear across to the states and provinces of the Rocky Mountain Cordillera. Indeed, they were members in good standing of what had come to be called the "Pacific Province." The Avalon fossils, plus some from isolated pockets along the Atlantic seaboard (St. John's area of Nova Scotia, the aforementioned Braintree slate in the Boston Basin, and some exposures only recently found to contain such fossils in North Carolina) belonged, naturally enough, to the "Atlantic Province." And that province is just as far-flung as its counterpart: the fossils of North America's thin little sliver of "Atlantic Province" agree very well with those in Wales and Scandinavia. It really was a puzzle, that Newfoundland situation. It was easy enough to see why Europe's fossils were different from the vast majority of North American trilobites (i.e., those of the "Pacific Province"); but why should the eastern-most exposures of Cambrian rocks in North America produce essentially European fossils?

Marshall Kay might have been expected to resist the notion of continental drift, or its modern version, "plate tectonics,"* as it caught on in the 1960s, reforming and revitalizing the entire science of geology. After all, he had for years been recognized as the inventor and champion of another theory of the Earth—one that did not conflict with drift so much as it employed utterly different terms. And Kay worked at Columbia, as recently as the 1950s a bastion of anti-drift feeling. (I had been taught that drift was nonsense as an undergraduate there in the early sixties; it was the "new truth" by the time I entered graduate school at the same institution.) But it was Columbia geologists as much as anybody else who established the validity of plate tectonics and wrote much of the revised theory of the Earth. And Marshall Kay helped turn the tide; for as many an old-line geologist before him, Kay knew in intimate detail how well the geology matched up on both sides of the North Atlantic. In fact, to the surprise of nearly everyone save himself, Kay became an ardent "drifter" toward the end of his career. And that is when he showed up with his little collection of trilobites from *central* Newfoundland.

Covered with virtually impenetrable shrub and teeming with voracious black flies to bedevil the hapless explorer, central Newfoundland is something of a paleontological nightmare, as difficult and unrewarding as the eastern and western flanks are bountiful in repaying visit after visit. Most reconnaissance is done by boat around the perimeter of the small islands off the northern coast, in Notre Dame Bay. New World Island is one such place on which

*The Earth's outer crust is divided into some eight major and numerous minor regions, or *plates*. The relative motions of these plates, plus the resulting change in continental and oceanic positions and the development of mountain systems, are known as *plate tectonics*.

Kay and his students chose to focus their attention. Just to the south, on Dunnage Island, in the so-called Dunnage melange (a mixture of badly deformed rocks, so confusing that I am told that Kay originally mapped it upside down), he found a thin band of limestone running from the shore out into the water. In the limestone were the telltale black blotches of trilobite remains—the first signs of the trilobites Kay was to ask me to identify.

Hampered though I was by a nearly total lack of experience, I did identify them, but not without difficulty. I used the trilobite bible—Part O of the *Treatise on Invertebrate Paleontology*—and identified them as *Bailiella* and *Kootenia*, two Middle Cambrian trilobite genera, even though no fossils older than Lower Ordovician had yet been recovered from Newfoundland's mid-region. At least the trilobites Kay had brought back *seemed* closest to those genera among all those described in the scientific literature so far—a conclusion thankfully confirmed by several Cambrian trilobite workers to whom I later turned. Both *Bailiella* and especially *Kootenia* are well known and fairly abundant sorts of trilobites. But there was one remarkable thing about them: *Bailiella* is from the Atlantic Province, while *Kootenia* hails from the Pacific Province.

Such "mixed" faunas had been found before, notably in the Taconic region of eastern New York and western Massachusetts and in other manifestly "disturbed" areas where the rocks had been thrust westward from some unknown source farther to the east. Here in Newfoundland there was no indication of any such long-distance rock travel. What did catch our attention, though, was the recently published suggestion that there had been a proto-Atlantic ocean in the Cambrian. Narrower than the present-day Atlantic, that ocean (since named "Iapetus," after one of the Titans) closed; and when the North Atlantic as we know it opened up (beginning around 100 million years ago), the fracture line did not quite match the old boundaries of Iapetus: part of Cambrian Europe stuck to North America to form those isolated pockets.

Kay had, it seemed, stumbled on a perfect piece of corroborative evidence: the limestone encasing those trilobites implied shallow water, most likely surrounding a mid-oceanic island. The two trilobite species that managed to get out there and populate those fringing waters just by coincidence had come from each side of Iapetus.

Or so it seemed to us. It is tough to reconstruct the world back then. But much can be done by just matching up fossils with pictures in books: whole ancient worlds take on a shape, and seem to reveal things undreamt of. Two years before that work, no one had ever given serious thought to the existence of a "proto-Atlantic" ocean.

Trilobites, Evolution, and Geologic Time

There are hundreds of genera of Cambrian trilobites. Many formations of the Cambrian are literally jammed with trilobite fragments. Trilobites, like crabs, periodically undergo molting: the hard exterior skeletons split open along programmed lines of weakness. Typically, the animal will take in extra water as part of an extremely rapid growth spurt in the hours between the shedding and the hardening of the new shell. Trilobites had many such lines of weakness—"sutures"—that divided the cephalon (head) into several parts. A major suture nearly always ran from the back side of the head, through the eye (separating the lens surface from the palpebrum, or "brow" section), and then around the front of the head. Another suture (sometimes two) then connected this "facial" suture with the underside of the head, where a "rostral" plate was generally surrounded by sutures. Another plate, the hypostome, was always present. The central region of the head, the "glabella," attached to the "cheeks" up to the eye region, constitutes the "cranidium." All this is merely to say that when trilobites molted, they usually did not simply crawl out of one escape hatch which would then snap shut, leaving a single discarded skeleton, as modern horseshoe crabs do. Instead, trilobites left a puzzle of parts, bits and

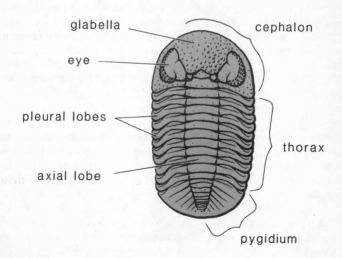

FIGURE 13   Sketch of trilobite anatomy.

pieces, especially in the Cambrian (before many trilobite groups simplified the molting process). Complete, entire trilobites are common enough in Cambrian rocks—but are in a decided minority.

It is still conventional to think of paleontology as *either* primarily a geological subject or a part of biology. Tradition has it that invertebrate paleontology is mainly concerned with geological topics, whereas vertebrate paleontology sprang more from the science of comparative anatomy. Vertebrate fossils, which with notable exceptions tend to be rare, cannot be counted upon to solve geological problems, such as the correlation of rocks. Invertebrates, on the other hand—and the billions of molted remains of trilobites cramming Cambrian rocks are certainly a case in point—are easily collected at thousands of localities the world over, and thus ideal material for deciphering the pages of Earth history.

Two isolated bodies of rock strata are "correlated" when someone demonstrates some special features common to them. Correlation usually has the stronger meaning of "the same age." Thus we no longer correlate rocks on the basis of their similar mineral content: the presence of the uranium mineral carnotite, for example, in various Mesozoic rocks of the western United States, cannot be used as a means of correlation. The standard, if not sole, criterion for correlation is the fossil content of rocks. And it is germane at this juncture to note that the principles of correlation were worked out nearly simultaneously by William Smith (1769-1839—the "Father of Stratigraphy"), looking at invertebrate fossils, and Baron Georges Cuvier (1769-1832) and Alexandre Brongniart (1770-1847), working on the vertebrate- and invertebrate-fossil-bearing strata of the Paris Basin.

Smith was particularly clear about his principles. Employed as a surveyor on the canal projects of England's Industrial Revolution in the waning days of the eighteenth century, he found that while climbing successive hills in a region along the right-of-way, he would observe similar sequences of fossils. He could go to local naturalists' home "cabinets" and tell them the relative order, from bottom to top, in which they had found their specimens. It was only becoming clear then that there was indeed a grand order to fossils the world over—or at least in Europe and North America, where informed amateurs were already poking around and learning quite a lot (Thomas Jefferson being an early and modestly distinguished example, describing some of the extinct Ice Age [Pleistocene] fossils of the still famous Big Bone Lick of Kentucky).

As we have seen, the basic idea of correlation with fossils is that the "same" fossils imply the "same" age. Similar fossils imply a more "iffy" match-up in time. Years ago, Darwin's "bulldog," Thomas

Henry Huxley (1825-1895), pointed out the perils in such assumptions: two dissimilar faunas may nonetheless have lived at the very same time, but simply in different regions, or in utterly dissimilar habitats. There were, for example, dinosaurs on Earth at the same time that ammonites and mosasaurs swam the seas over the continental interiors. By the same token, Huxley warned, we could mistakenly claim that two very similar faunas implied the same age when actually they were millions of years apart. I can testify to that: Upper Silurian trilobites, brachiopods, clams, and snails look very much like those living in comparable environmental conditions in the Lower Devonian—a total span of some 25 or 30 million years. The problem was especially acute back in Huxley's time; today, detailed study of these fossils permits a handful of experienced old hands (I am one) to tell at a glance where in the Upper Silurian or Lower Devonian a particular fossil comes from. The old hands are usually right.

Yet on the whole the correlation system works. And that is simply because life has evolved. It has had a history. Belgian paleontologist Louis Dollo (1857-1931), whose career spanned the turn of the last century, gave us "Dollo's Law," which said that evolution is irreversible. He was making a generalization (which is exactly what all scientific laws are: codifications of regularities in nature). In general, we can be pretty confident that exactly the "same" extinct animal or plant species will never reappear: though some later species may be confusingly similar, in no case will it be the same species. Dollo was dealing with a special case: the loss of advanced features may make some future descendant confusingly like some remote progenitor. However, he encouraged us to take heart: it is unlikely to the point of virtual impossibility that such a reversal will end up in a descendant utterly the same as its ancestral form *in all respects*.

Thus evolution gives a signal of *change* to the fossil record. The pattern of that change is a matter for later discussion: all we need know now is that evolution will change the composition of the creatures alive, the organisms that produce our fossil record, as the ages roll by. As sporadic and erratic as the rate and direction of change can be, it is the natural process underlying the observations of Smith and of Cuvier and Brongniart.

But ironically, sometimes when the fossil record is especially rich we can be deluded into thinking that the kind of patterns observed by William Smith in his Jurassic ammonites (ancient molluscan relatives of the pearly nautilus) are typical of *all* situations in the fossil record. Cambrian trilobites are another example of such paleontological riches. Fossils in such circumstances seem to be laid out in simple belts ringing the globe. The old Latin word for belt is *zona*, and that's what geologists began calling bodies of rock characterized by a typical suite of fossils. It began to be expected that such "zones" bearing a rigidly characteristic suite of fossils would pop up, always in the same order, the world over. It turns out, though, that the fossil record is not usually laid out in a sequence of easily detected zones.

The problem is that not all species appear lockstep at the same time. And not all of them disappear at the same time. There is a normal, clockwork ticking that involves the introduction of new species (either by evolution or, more usually, by migration into a region). There is, too, a normal rate of extinction that takes out one species, then another, in a statistically regular sort of way. Life is not ruled strictly by episodes of nearly total extinction, followed by periods of simultaneous re-proliferation as life expands once again.

But whenever there *are* such regular waves of extinction and proliferation, following each other with a regularity of, say, a few million years; and whenever such events are sufficiently severe as to be fairly worldwide (if not utterly global), then we will get those easily seen "zones" that seem to follow in the same sequence the world over. Correlation is a lot easier when such clearly laid out zones are formed; it is a lot trickier when different species appear and disappear sporadically.

All of which takes us back to the Cambrian. Zones are pretty clear in the Middle Cambrian—and especially the Upper Cambrian. It seems at the outset that we are dealing with some sort of extinction/proliferation cycle. Since more than one group of organisms is usually involved, we might see the problem as fundamentally ecological in nature. But it is also evolutionary in a most fundamental sense: for we are dealing with the ecological controls over evolutionary events; the origins and extinctions of real species. The importance of evolutionary and ecological patterns in forming the fossil zones crucial to geological correlation is nowhere more evident than in the Cambrian, as we shall now see.

Biomeres: Trilobite Zones and Evolutionary Patterns

Once again we go to Utah, this time to its west-central reaches as U.S. 6-50 skirts the ranges and heads toward Nevada. Where the intermittent Sevier River crosses the highway, the little town of Delta stands as the lone outpost of civilization. Yet even Delta seems dated, if by no means as antiquated as its surrounding rocks and fossils: wives can still be seen patiently sitting alone in pickup trucks while their husbands frequent one of Delta's two bars, and kids still roar up and down the main drag in Chevies.

Delta is the jumping-off point for the House Range. A few miles beyond the town, a dirt track leads off toward those distant mountains, cutting across dry, scrubby sheep and coyote rangeland. Drawing nearer to the mountains, the track diverges; to the right lies magnificent collecting in Middle Cambrian rocks. Here is the famous Wheeler Ampitheater at Antelope Springs, with its gray Wheeler Shale that has yielded countless thousands of complete trilobites. One of these trilobite species, *Elrathia kingii*, is among the most familiar of all fossils, not just here in the United States, but over much of the world. It has been quarried literally by the truckload from this remote spot nestled up against the House Range, and specimens adorning tie tacks, cufflinks, necklaces, bracelets—or just serving as knick-knacky paperweights—are to be found in virtually every rock shop from Maine to California. And *Elrathia* is but the most common species of a richly varied trilobite fauna, all jet-black, in distinct contrast to the light gray matrix in which they abound.

But it is to the left as the track splits that we find Upper Cambrian rocks, and they too are jammed with trilobites. In contrast to the trophy-like complete specimens regularly coming from the older Wheeler formation nearby, these dense Upper Cambrian limestones yield only those fragmented bits and pieces left after trilobites either molted or died and left their outer skeletons to be worried and winnowed by the agitating currents and waves of those warm, shallow Upper Cambrian seas. Here, in places like the House Range and other localities in Utah and Nevada, paleontologist A. R. "Pete" Palmer discovered what seemed like a new wrinkle in the old story of fossil zones nicely stacked atop one another. Following it through, we begin with an insight into geology, and watch it transform our appreciation of how evolution has regularly worked to form the history of life.

The Upper Cambrian is widespread in North America; the Lower Cambrian is restricted entirely to the margins of old North America, lapping up on the old continental shelves and cropping out now only in odd places in the Appalachians and southeastern California and southwestern Nevada. Seas of Middle Cambrian times penetrated the continent only a little further; nowadays, we find Middle Cambrian rocks and fossils throughout the western Cordillera, but not in the central regions of the continent. But in the Upper Cambrian, the seas finally did encroach on the continental interior for the first time since the advent of complex life. The Wisconsin "dells," along the upper reaches of the Mississippi River, are buff-

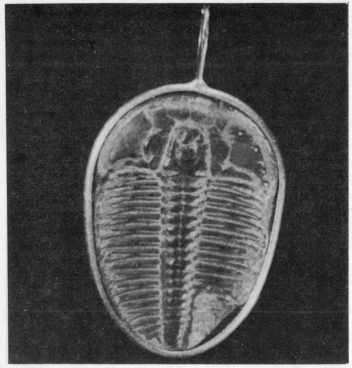

FIGURE 14 North America's most peripatetic fossil, the Middle Cambrian trilobite *Elrathia kingii*. Thousands of specimens collected at the Wheeler Amphitheater in the House Range of Utah have been distributed to souvenir and craft dealers around the world. This one has been made up into a pendant.

colored sandstones and siltstones bearing a characteristic Upper Cambrian fauna dominated by trilobites—a fauna first studied in detail by the pioneer American paleontologist James Hall. Similar fossils occur up and down the Appalachians, and in the Ozarks, the Llano region of Texas, the Arbuckles of Oklahoma, and the Rockies of Montana. With so many scattered regions of outcrop, the task of correlation—of deciding how these far-flung rocks match up in time—looms as an important piece of business. This was Palmer's major objective as he climbed the Upper Cambrian rocks of the Great Basin, measuring their thickness and collecting their trilobites.

Trilobites really were so common in the Cambrian, and so prolific in their diversity, that Upper Cambrian rocks seem to be regular sequences of successive "zones." The *Aphelaspis* zone (characteristic trilobites lend their names to each zone) might be hundreds of feet thick in one range and only a few tens of feet thick elsewhere. And as William Smith observed all those years ago with his Jurassic rocks and fossils, zones always appear in the same order: the *Aphelaspis* zone occurs above the *Crepicephalus* zone, and immediately below the *Dicanthopyge* zone. Thus the job is straightforward: go to the unexplored ranges, such as those of the Great Basin of the West; collect the trilobites; identify them, and describe them in the technical literature; and, on the basis of their similarity to trilobites elsewhere, match up the sequence of zones in the Great Basin with zones known elsewhere.

But the Great Basin is itself a huge area, and Palmer first had to match up the trilobite sequences from each of the ranges. And that's when he first noticed something strange. It is always assumed that a zone is, at least as far as anyone can judge, a band of rock representing a chunk of time. Find that zone anywhere, and you are within that time span. This means that it is assumed that the zones began everywhere at precisely the same time, and ended precisely at the same time as well. Even if we worry that species characteristically originate in a fairly localized area and, if found to occur over truly broad regions of the Earth, must have actually spread after their first appearance, our concern is allayed as we realize that zones over 500 million years old themselves represent at least several millions of years: our time-telling is so coarse that we cannot hope to document the spread of a species, or an entire fauna, from one region to another. Within the limits of our resoltuion, there will be plenty of time for such migration—and it will all *look* as though it had been simultaneous anyway.

But there are so many good exposures of Upper Cambrian rocks in the Great Basin that Palmer was indeed able to document migration. Or so he interpreted the rather odd pattern that emerged from his data. For Palmer seemed to see two sets of overlapping zones with his western trilobites. There was the usual sequence, long familiar to him. And there was a series of larger chunks, each containing several zones—in and of itself no great surprise, as paleontologists had for years been accustomed to grouping zones into larger units called *stages*. But the peculiar thing about Palmer's data is that the bottoms and tops of his larger units, which he called *biomeres* seemed at odds with and actually to have cut across, the bottoms and tops of the lower and upper zones (respectively) within those biomeres. Palmer had two conflicting classifications of the same rocks based on their fossil content.

Clearly, or so it seemed to Palmer, one or the other unit was *time-transgressive*: instead of being about the same age everywhere, one of the units was younger in some places than it was in others. Keeping the old assumption that the zones themselves have had synchronous beginnings and ends, Palmer published his notion of "biomeres" as biostratigraphic units the size of a stage, but with boundaries that cut across time—being younger in the western than the eastern region of the Basin and Range. There was an immediate uproar—many paleontologists simply claimed that biomeres *are* stages; indeed, most recent discussions of biomeres deliberately downplay the time-transgressive nature of the boundaries. Attention has shifted to what biomeres really are: evolutionarily coherent faunas with a characteristic beginning, history, and end, followed in an almost rhythmic pattern by another wave: the clock is reset, the pattern begins anew, and history is played out along similar—though never identical—lines again.

It is worth a closer look at the internal workings of these biomeres, as they reveal some very basic aspects of the pace of stability and change utterly characteristic of the entire history of life—whether the organisms involved are Cambrian trilobites, Mesozoic ammonites, or Tertiary mammals.

Palmer thought he was seeing migration. Back in the Upper Cambrian, Nevada and Utah were perched on the very edge of the continent. Careful study of the rocks from west to east reveals the presence of much deeper water, well down the old continental slope in the western reaches of the Great Basin. By the time we reach west-central Utah, rocks of the same age are composed of limy sands, hallmarks of much shallower waters. It looked to Palmer as if periodically deep-dwelling trilobites would appear, invading the inner, shallower reaches up over the continent, and in so doing radiate into an array of differently adapted species. The species would live on for a time; indeed, several zones would come and go. Then, very suddenly, the entire ecological system would disappear—to be replaced by a new stock of trilobites from the deep water penetrating the shallower waters up slope. Only now, as the rhythm begins to come clear, can we see the invasion of the generalized deep-dwellers as perhaps yet another example of life taking advantage of an opportunity long closed off: it would seem likely that the deep-dwellers' invasion of shallow waters occurred only after the well-entrenched, diversified trilobites already there had suffered extinction.

There is a fascinating comingling of genealogy and ecology in the biomere story. Palmer named each biomere (he initially found three, although since then other paleontologists have found more, and indeed have documented them far away from the Great Basin) after an entire family of trilobites. Thus the "Pterocephaliid biomere" was

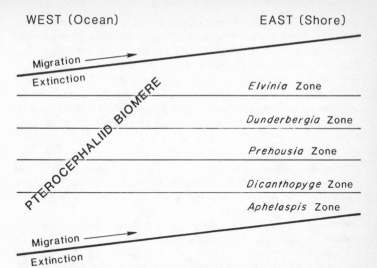

FIGURE 15 Zones of Palmer's pterocephaliid biomere of the Great Basin, Utah and Nevada. Slanted lines indicate Palmer's conclusion that the tops and bottoms of biomeres reflect episodes of extinction, followed by repopulation from the open ocean. Because both extinction and immigration affected the western regions before reaching the eastern, shoreward areas, biomere boundaries transect time, becoming younger as one travels from west to east.

named for the Pterocephaliidae, a family of trilobites whose name means "wing headed" (there being a rather broad fringe on the front of the head of *Pterocephalia* and some of its closer relatives). And indeed, the radiation into different niches once the ancestral stock invaded the shallower environments seemed repeatedly (i.e., in biomere after biomere) to involve predominantly members of a single lineage, all derived from the same common ancestral species. This is the purely evolutionary, genealogical side to biomeres: they record adaptive radiations of entire stocks of ancient organisms.

But biomeres include many different kinds of organisms in addition to those forming a single evolutionary family. Species of trilobites unrelated to those whose family lends its name to a particular biomere, as well as brachiopods, echinoderms, and other forms of Cambrian life were all there as well. Thus biomeres give us two things: the opportunity to witness the origination and fate of new ecological associations as well as the chance to analyze the explosive radiation of single stocks of organisms—large clusters of related species all descended from a single common ancestor.

Palmer's biomeres forcefully remind me of the situation some 100 million years later in which a fauna of clams, snails, brachiopods, and corals is dominated by a spectacular radiation of trilobites in the southern hemisphere supercontinent of Gondwana. Here, in the Lower Devonian, now exposed in the Andes and the lowlands of the Amazon, as well as the mountains of Antarctica and South Africa, we have the remains of some forty different trilobite genera. Of this rather prodigious array (representing a time span of some 20 to 25 million years), fully 60 percent belong to one family, the Calmoniidae, which underwent an explosive evolutionary development.

Whether or not the invasion from deep to shallow water went slowly enough to register a time-transgressive effect in the rocks remains debatable, and in any case it is not the main item of debate these days. Palmer's biomeres show that historical ecological packages can be understood as systems with their own beginnings, histories, and eventual terminations. The history of an ecological system is intricately interrelated with the histories of separate genealogical lineages. What terminated those ecological systems that formed the biomeres is still an active bone of contention: some say a drastic drop in temperature eradicated the shallow-living, rather more specialized species, while others claim that the rocks show a sudden lowering of sea level that would simply have reduced, or eliminated, the shallow water habitat. Whatever the precise environmental cause of the sudden habitat change, when the shallow waters once again became habitable, a different bunch of colonists got in and established similar ecosystems that proceeded to undergo similar histories and eventual fates—similar, but not identical. We will be encountering these themes over and over again throughout the history of life.

Cambrian and Later Paleozoic Interface: Deciphering Gaps in the Fossil Record

Oslo, Norway, to American eyes, is a curious blend of the strange and the familiar. The granite hills above the city provide the foothold for a mixed hardwood and evergreen forest very like the woods of northeastern North America. On the other hand, you can read a newspaper outdoors at 2 A.M. on June 21 yet still feel you are in a temperate clime, and some of the country houses have sod-and-grass roofs. What is unique about Oslo is its magnificent fjord, dotted with islands; to a paleontologist, the Oslo fjord is a goldmine of Cambrian and Ordovician fossils, especially compelling to a trilobite lover.

Trilobite history is a bit at odds with the rest of the organic record. These animals dominated life in the Cambrian; they were fairly prolific in the Ordovician, the next succeeding geological period, but they were always out of step with the major events affecting, nearly simultaneously, the fates of entire cadres of other organisms. There was, for example, a vast explosion of hard-shelled invertebrate organisms in the Lower Middle Ordovician that hardly affected trilobites at all. Trilobitan vicissitudes, both their triumphs and their disasters, are ill-correlated with those of other groups. And the islands of the Oslo fjord offer a good place to observe this frozen action.

At the very top of the Cambrian sequence in Wales lie rocks of the Tremadocian series. (*Series* are subdivisions of *systems*; the Cambrian, Ordovician, Silurian, etc. are systems of rocks, corresponding to *periods* of time with the same name.) The British consider the Tremadocian as part of the Cambrian: the trilobites and other fossils for the most part sport a Cambrian look, and the next succeeding strata, the Arenig sands, seem physically more closely related to the overlying Ordovician rocks. There is generally a disconformity indicating a gap in recorded time, between Tremadocian and Arenig rocks. Yet in North America, the Tremadocian seems to fit in better with what we are accustomed to classify as Ordovician rocks.

The Tremadocian is well developed in the Oslo fjord. Trilobites are plentiful, but to the discerning collector, also a bit puzzling. For it is hard to look at many of these Tremadocian trilobites and say, with anything approaching conviction, that it belongs to a well known family *outside* the Tremadocian—whether above or below. Tremadocian trilobites are something of a breed unto themselves.

Cambrian trilobites (and Cambrian trilobite workers, for that matter) are notoriously considered (among paleontologists, at least) breeds apart from their "post-Cambrian" counterparts. One would think that the Tremadocian would smooth the problem over, allowing us to see the connections between trilobites of the Cambrian and those of successive Paleozoic periods. For the most part it doesn't, and we really must ask why. Nor is it a matter of paleontologists focusing on one section of the geological column while failing to examine trilobites of different periods, as is sometimes said. When I was looking at Tremadocian trilobites in the Oslo fjord, I was with a raft of other "trilobitologists." Altogether we covered the entire spectrum of trilobite history; we spent the day island hopping and collected trilobites from the Middle Cambrian up through the Middle Ordovician. By the end of the day, the Tremadocian seemed as much a world apart—at least as far as its trilobites were concerned—as it had when first we jumped aboard our little ferry.

The fossil record is full of these discordances—these non-connections between representatives of what we know *must* be a single, coherent group. Trilobites, after all, share a common body plan unlike that of any other group of arthropods. We have no trouble seeing that as a whole they are an evolutionarily coherent group; but we have no end of problems trying to figure out how all the various subgroups of trilobites are interrelated. The biggest trouble lies at the interface of the Cambrian and later Paleozoic, where the complexion of trilobite life seems to have changed so radically. And, frustratingly, the intervening Tremadocian, when it is found at all, tends to add to the confusion rather than give us those "missing links" that it would be so comforting to have.

Another example points the way to understanding why such difficulties constantly arise in trying to decipher evolutionary

FIGURE 16   The Ordovician trilobite *Triarthrus*. Olenids such as *Triarthrus* are holdovers from the Upper Cambrian.

histories in the fossil record. The blastoids are a group of echinoderms which, like the crinoids, lived attached by a long stalk to the sea bottom. First appearing in the Ordovician, they were always a relatively minor part of seafloor life until Lower Carboniferous times (what we call the Mississippian Period in North America—a part of the early Upper Paleozoic, beginning around 360 million years ago). In the Mississippian, blastoids simply went wild. You can collect them by the bushel in some horizons in the Mississippi Valley. Yet in the overlying Pennsylvanian rocks, not a blastoid is to be found—nor were they known from any younger rocks—not, that is, until they showed up in the Permian (uppermost Paleozoic) rocks of the remote Indonesian island of Timor, along the outer face of a volcanic island arc. More blastoids have since turned up in other Permian rocks, but not, at least so far, in the Pennsylvanian. Nonetheless, though we haven't found their fossils, they must have been in existence in the Pennsylvanian; indeed,

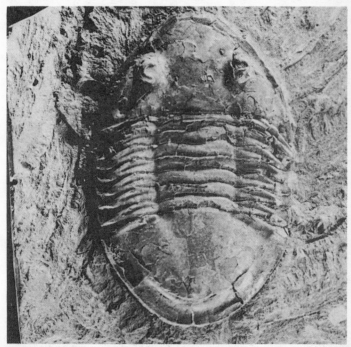

FIGURE 17 The asaphid trilobite *Isotelus* represents a group that first appeared in Ordovician times, with no obvious Cambrian relatives. This three-inch-long specimen is a relative baby—isotelids over two feet long have been collected!

many of the creatures living in Pennsylvanian times seem very similar to those alive during the Mississippian, making it hard indeed to see why the blastoids should have disappeared at all. Yet disappear they did. Of course, it is even harder to believe that blastoids really did become extinct at the end of the Mississippian only to have re-evolved from some other echinoderm as the Permian began.

What is the answer? How are we to explain these great gaps in the fossil record? Recall Palmer's primitive stocks of trilobites living in the cold waters down the continental slope of the western United Stated during Cambrian times. What is fortunate about that story is the actual knowledge we have of those down-slope faunas, recovered from rocks in the western sector of the Great Basin. Recall, too, the serendipitous discovery of what seems to have been a mid-oceanic island habitat, now a thin string of limestone on a remote Newfoundland island. The picture becomes clearer, and though its message is sobering, at least it helps us appreciate the very basic nature of the fossil record. The plain truth is that, except in very rare circumstances, most of the fossils we have come from sea bottom sediments *deposited on the continental surfaces*. The consensus is that the Earth's diameter has remained relatively constant, at least throughout the last 600 million years (though there is a minority of opinion that holds that the Earth has been expanding). The continents back then were, if anything, a bit smaller than they are now (remember the continental slope's position in western North America during Cambrian times), so we can safely conclude that the true oceans, (as opposed to the thin veneer of seawater that was sometimes present atop the continental surfaces) were at least as large as they are now.

Now, one of the more astonishing facts about the oceans that has emerged over the past twenty-five years of intense exploration is that nowhere are the oceans older than Middle Jurassic, or some 175 million years old. Nor is this a matter of inadequate sampling: new crustal material is constantly generated along spreading ridges (such as the Mid-Atlantic Ridge) and as this molten material spreads laterally, cools, congeals, and sinks, entire crustal blocks (the famous "crustal plates") are shoved along too. Thus the older oceanic rocks should be at the outer edges of the plates, and that is precisely where we find them.

But the Earth is supposed to have a constant size. If new crust is being created, and if indeed as must be so, there were oceans in the pre-Jurassic world, where is that older oceanic crust? Gone. Where opposing plates are in collision, oceanic crust is often swept down below continental crust. This "subduction" (going on right now, for example, as the Indian Plate slides under the Asian Plate along the

Indonesian outer archipelago) systematically destroys the leading edges of plates exactly as fast as new crustal material is produced by the upwelling of molten rock at the back edges of these same plates—an equation that leaves the Earth the same size throughout geologic time, and explains why we have next to no hope of ever finding true open-ocean fossil records going back beyond 175 million years regardless of where we look—whether on the continents or in the oceans themselves. Only in those mangled slivers of oceanic crust that were miraculously spared total subducted destruction, and have now been shoved up and incorporated into moutain belts whose ancient roots are finally exposed by erosion, can we by pure chance stumble on oceanic fossils of truly ancient age.

If such an evaluation paints a grim picture of the fossil record's completeness, just think how truly lucky we are (in more ways than one) that these are indeed unusual times in the Earth's history: the normal state of affairs is smaller polar ice caps, and hence more water—which in turn spills over the edges of the oceanic tubs and floods the continents. Had that water been locked up during the Paleozoic as it is today, marine muds would not have accumulated on the continents to anything near the extent that they did. As a result, we'd have next to no fossil record, at least of the marine creatures that, as things actually stand, form roughly 99 percent of that record. On the other hand, were we not still basically in the throes of a period of pronounced episodic glaciation (albeit currently enjoying something of a respite from the ice), we would probably not be here to ponder the fossil record at all.

What this all adds up to is a simple recognition that in the nature of things, we will never see many of the creatures that lived in those bygone seas. Indeed, the job of sampling everything alive today is difficult enough, an endeavor still incomplete and likely to remain so, as the extinction of modern species outstrips our ability to record them before they pass along. The frustrating gaps in trilobite history—gaps mirrored in every group that has left a fossil record—are unlikely to go away simply with more collecting, unless we get lucky and find more truly oceanic fossils of the appropriate age.

Yet the situation is far from hopeless. We need only remind ourselves that there is one grand, single pattern of resemblance linking all forms of life. Trilobites as a whole form a coherent group, sharing a common plan of body organization not seen in any other arthropod group. Despite those vexatious gaps, and the likelihood that for all our future efforts at collecting, those gaps are destined to remain unfilled, we still can frame a comprehensive, and accurate, picture of how all the various forms of trilobites are interrelated.

Recent advances in the analysis of evolutionary history ("cladistics") have capitalized on the fact that the grand pattern of resemblance linking all of life at any one moment can be discerned regardless of how many intermediate forms along the way have become extinct and remain undiscovered in the fossil record. We can tell that whales are mammals (with their four-chambered hearts, mammary glands, and other features typical only of mammals), even though there is still little by way of good intermediates between primitive mammals and true whales. And so it is with trilobites: with a judicious mix of field collection, laboratory cleaning and examination, and sharpened methods of analysis, we are now finally seeing connections emerge between trilobite groups of the Upper Cambrian, Tremadocian, and Lower Ordovician.

Later Trilobites

I keep hoping someone will dredge up a living trilobite. We'd know so much more about their anatomy, physiology, and behavior—the basic *biology*—of these creatures from a single living specimen than we can ever divine from thousands of fossilized specimens drawn from hundreds of different species. When Dr. Jacob Green, a Philadelphia physician, wrote the first real monograph on North American trilobites (in 1835, with a supplement in 1838), he came to the place where he had intended to speculate on the zoological affinities of trilobites—but announced that owing to the recent recovery of living specimens from the deep sea, he would sit back with the rest of the world and await the scientific description of the creatures. What a shame—those "trilobites" turned out to be marine isopods, a group of crustaceans that also includes garden pill bugs. As much as these serolid isopods really do resemble trilobites, they have the jaws, double set of antennae, and other typical features of crustaceans; serolids simply are *not* trilobites. Trilobites seem to have gone forever.

But, before they left, they compiled a pretty impressive history—the entire Paleozoic, bottom to top, in a sense is defined by the longevity of the trilobite clan. The youngest trilobites, from two different species, appear to be those collected from well up the face of Mt. Everest, from Upper Permian rocks. And it is fitting that they are proetid trilobites—the closest thing to "standard issue" as far as trilobites go. And therein lies another generalization about life's history: what has been called the "survival of the unspecialized."

Proetid trilobites first appeared in the early Ordovician. Though during the long haul from the base of the Ordovician to the end of

FIGURE 18 The Carboniferous trilobite *Griffithides*, a proetid, and member of the only trilobite stock to survive the Upper Devonian.

the Permian—some 260 million years—many proetid species came and went, and they even experienced occasional evolutionary bursts that produced some fairly novel-looking species, on the whole proetids were a remarkably stable lot. Proetids are one group that has recently yielded to concerted efforts to link them with Cambrian trilobites. Thus their phyletic pedigree is indeed old.

Why should primitive, relatively anatomically generalized forms tend to hang on, outliving their more specialized offshoot relatives? After all, we are accustomed to thinking of evolution as a matter of building ever more refined, efficient mousetraps—shorthand for saying that anatomical advancement, leading very often to increased specialization, is an inevitability through the course of time. And indeed we see this increased specialization time after time. But, as paleontologist Henry Shaler Williams was pointing out before the turn of the century, (and, more recently, as Peter Bretsky has stressed), the more specialized a species is the less able to cope with change a species will be once the inevitable happens and old habitats change beyond the point of recognition. Anatomically primitive organisms tend to be ecologically more generalized than their close but more specialized relatives. It simply stands to reason that of the two, it will be the jack-of-all trades that has the slightly improved

FIGURE 19 The phacopid trilobite *Phacops milleri*.

statistical chance of resisting extinction. When mass extinctions come *nearly everything* goes; less severe episodes, though, allow some winnowing, and this is when the survival-of-the-less-specialized rule seems to apply.

Proetids in the Ordovician were just one of a number of different trilobite groups, which also included phacopids, odontopleurids, lichids, asaphids, calymenids, olenids (left over from Cambrian times, as were the agnostids), illaenids, trinucleids, and a number of lesser groups. Together they formed a rich and varied lot, and most Ordovician faunas readily produce specimens from five or ten genera—and sometimes more. At the end of the Ordovician, extinction claimed many groups. It's as if trilobites, after the Upper Cambrian, suffered a steady reduction in the numbers of their major subdivisions—if we qualify that word "steady" to mean episodes of reduction followed by long periods when all surviving groups retained a very vigorous ecological presence, and even expanded their numbers through evolving additional species. Yet after the Ordovician appearances, no major new groups of trilobites ever appeared again.

The Upper Devonian extinctions, among the most severe in the history of life, took out all but the proetids. After that brush with total annihilation, trilobites, via their sole, proetid relatives, under-

FIGURE 20   A *Phacops* lookalike, the calmoniid trilobite *Bouleia dagincourti* from the Lower Devonian of Bolivia. Calmoniids are distant cousins of phacopids.

went some diversification in the Lower Carboniferous (Mississippian). But when the massive end-of-Permian crisis gripped all of life on a worldwide scale, it was really no epochal disaster for trilobites, for they had long since dwindled to but a few species. True, had these last survivors managed to make it through that Permian wave of extinction, trilobite genetic heritage would have been preserved, and possibly trilobites once again would have recovered, diversified, and become a force in marine ecosystems throughout the world. But 325 million years of existence is, after all, not a bad track record.

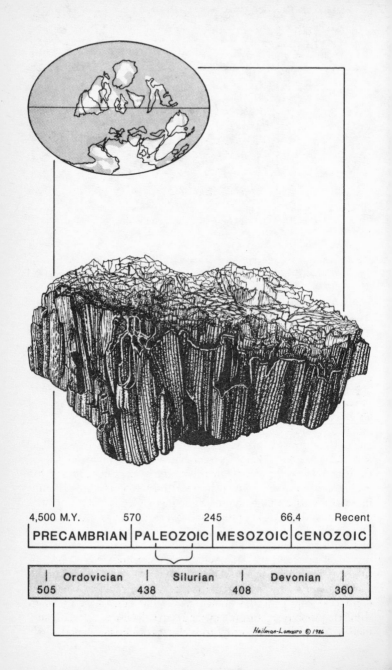

4,500 M.Y.   570   245   66.4   Recent
| PRECAMBRIAN | PALEOZOIC | MESOZOIC | CENOZOIC |

| Ordovician | Silurian | Devonian |
505   438   408   360

Heilman-Lomauro © 1986

4

Life in Paleozoic Seas

Lurking somewhere in the meaning of "evolution" is the idea of "progress." The Greeks philosophized about progress, and although the concept was seemingly (and understandably) all but dropped through the Middle Ages, progress as an idea enjoyed its greatest popularity as the Industrial Revolution began to pay off in material dividends. Change is tolerable—even desirable—if it is for the better. It was in precisely this climate that Darwin conceived his evolutionary ideas and wrote his *On the Origin of Species* (1859). And if Darwin himself was circumspect about completely linking organic evolution and the idea of progress, others were not as careful: social Darwinism, for example (which was not, in any case, an invention of Charles Darwin), saw competition for resources between individuals in a society as inevitable. According to this line of thought, it is a dog-eat-dog world, and we must accept it as such; we should celebrate the "survival of the fittest," as it provides a means of constantly improving the human stock. But even if we prefer to think that we can organize our own human behavior in the economic marketplace without resorting to such pseudo-jungle scenarios, we

79

still wonder whether competition in nature ensures change as inevitable. Is progress truly built into the evolutionary system?

Darwin was writing for a world that saw nature—especially biologic nature—as static. He emerged convincing himself and most of his readers that change is inevitable, and that change is indeed closely linked to the idea of improvement—of slow, steady progress. But on the face of it there is a seeming contradiction: if evolution moves from the simple to the complex, from the primitive to the advanced, and if descendants are truly superior to their ancestors, why then do we see so many apparently "primitive" organisms around? If the fossil record bears out our prediction, and bacteria are the first fossils to appear, why were they not totally replaced by the single-celled eukaryotes that came along 2 billion years later? For that matter, why didn't those primitive eukaryotic cells disappear once complex life truly got going in the basal Cambrian? Follow this line of thought out, and one begins to wonder why any organisms save mammals and flowering plants exist—and even there we might wonder why opossums, platypuses, and even shrews should have survived when so many other, more advancced mammals, have appeared. Taken to its extreme, this line of reasoning becomes absurd, focused as it is strictly along genealogical lines, taking no heed of the diversity of ways and means of making a living that exists for the organic world.

Until recently, most paleontologists who took an active, professional interest in evolution specialized in the vertebrates. And here, at first glance, the fossil record seems to behave beautifully: "fish" show up in the Ordovician, jawed fish in the Upper Silurian, amphibians in the Devonian, "reptiles" (I'll explain the use of the quotation marks later) in the Upper Carboniferous (Pennsylvanian), mammals in the Upper Triassic, and birds in the Jurassic. Even better, for a view of evolution-as-progress, is seeing the Devonian as "The Age of Fish," the Upper Paleozoic as a "World of Amphibia," directly succeeded by a "World of Primitive Reptilia," to be replaced by the great Mesozoic "Age of Reptiles" (especially, though not entirely, dinosaurs). Finally, the Tertiary is the "Age of Mammals"; no one has yet seen fit to designate any particular segment of geologic time for the birds. It is particularly in such simple-minded formulations as these that the imagined hand of progress leaves its greatest imprint. In fact, to take but one example, the mammalian life of the Tertiary does not represent "progress" over reptilian-dominated land life of the Mesozoic: mammals appeared fully 140 million years before the "Age of Mammals" ever got going—living, as Stephen Jay Gould has put it, in the interstices of the dinosaurs' world.

Contemplating a career as an evolutionary-minded paleontologist, and mindful of the vast riches of the fossil record of marine invertebrates and the importance that adequate sampling plays in evolutionary studies of modern creatures, I was dismayed at first by the almost *anti*-evolution aura of the rocky record of invertebrates. There was no real trace of progressive development at all, certainly nothing on a par with what seemed to me at the time to be an accurate rendering of vertebrate history. To be sure, discoveries of simple bacteria and algae in Precambrian rocks had already begun, but the near synchronous appearance of so many groups in the early Cambrian, and the final explosion that rounded out the panoply of marine life in the Lower Middle Ordovician, seemed just the opposite of a simple picture of progressive change in the evolutionary episodes in marine life for the remaining 475 million years up to the present. In this fear I was totally mistaken, but to realize the error required unlearning some rather basic suppositions about the nature of the evolutionary process—most particularly this pesky expectation of some rudimentary form of progress to come shining through the history of life.

It is in this vast stretch of mid-Paleozoic time that I think it is easiest to see the typical rhythm that evolution takes on. This phenomenon is strongly affected by a question of scale—and once again we see how important it really is to have a fossil record greatly condensing history to the point where a human being can deal with the larger-scale units, such as species, that play such important roles in history. Mid-Paleozoic rocks supplied the first example used to formulate the theory of "punctuated equilibria."[*] Punctuated equilibria is concerned with evolutionary phenomena in the mid-range scale: the minutiae of population genetics, in which change in gene frequencies within populations is studied on a generation-by-generation basis, occur on too fine a scale to be within the ken of paleontologists. The theory of punctuated equilibria focuses on patterns of anatomical change throughout the long history of a species, and examines as well the nature of change as one species gives rise to another. Stemming from this theory is the possibility that within separate lineages there may be processes that bias the births and deaths of entire species—a sort of higher level analogue of natural selection. It is with the gross aspects of a species' history, and what happens as new species are derived from ancestral species, that the fossil record begins to come into its own, providing a focus simply unavailable to biologists stuck in a single time frame.

[*] See my book *Time Frames* (New York: Simon & Schuster, 1985) for a detailed account of the content and history of the notion of punctuated equilibria.

When Gould and I first proposed the theory of punctuated equilibria back in the early 1970s, we stressed something that had been known to Darwin and, indeed, to all paleontologists who were contemporaries of Darwin: once they appear, species tend not to change very much at all. They may last 5 or 10 million years—sometimes even longer—and yet, while a very few might undergo the sort of gradual, "progressive" modification we have come to expect of evolution, most will stay pretty much as they were when they first evolved. This stability, or *stasis* in technical jargon, more than likely simply reflects continued habitat "recognition"—there being no reason to build a better mousetrap if the old one remains serviceable.

But the mid-Paleozoic shows us that the individual instances of species-stability "punctuated" by occasional bursts of speciation (the central pattern of evolutionary stability and change addressed by the theory of punctuated equilibria) have immediate consequences for the ecological organization, and thus the ecological history, of life. Entire assemblages of organisms remain recognizable for millions of years, with some species dropping out, then perhaps reappearing (through changes of locale, rather than through any violation of Dollo's Law). The assemblages remain recognizable because the component species contributing populations to those ecosystems have themselves changed little, if at all. Stability on an ecological scale reflects the stability of the component species—or is it the other way around? Do we have economic stability because the basic packages of genetic information—species—remain stable? Or do the species remain stable because their environments—and therefore the ecosystems—are stable? My guess leans more to the latter, though the question has only begun to be asked and the answers to these very basic questions posed by the fossil record are by no means clear.

The basic pulse of evolutionary life in the Paleozoic, then, is economic business-as-usual for intervals as long as 5, or even 10, million years. Ecosystems are established and tend to run on virtually unchanged until an extinction event disrupts the system. If the event is severe enough, many species will become extinct. After relatively mild extinctions, replacement ecosystems tend to resemble those they have replaced, simply because close relatives are still available to stock the habitat. More severe extinctions mean more radical change in the complexion of replacement ecosystems. Detailed charts of geologic time readily bear out this phenomenon: the names of the geological systems and the corresponding periods of time ("Ordovician," "Silurian," "Devonian," etc.) reflect the experiences of early geologists who saw natural divisions in life forms in the different bodies of strata. Each system, moreover, is divided into a sequence of series, in which the faunal turnover is relatively less severe. Major extinctions are less common than milder events—a fact that enabled early geologists to divide up the time scale in the first place. This pattern of periodic disruption of ecosystems in varying degrees of intensity is the dominant signal of life's evolutionary pulse—and nowhere is it better seen than in the ecosystems of the Middle and Upper Paleozoic.

Enter the Actors: Brachiopods and Clams

There really is something strange about the Cambrian. It is very much as if complex life got going and somehow, somewhere, someone decided that it wasn't quite right. But rather than starting all over again, a simple, radical, reorganization was effected: new players were recruited, to be sure, and old ones dropped, but there was also a matter of emphasis—an expanded role for some, a diminished role for others—all rather reminiscent of cabinet changes at the beginning of a second presidential term.

Trilobites, as we have seen, played a diminished role after the Tremadocian. So did archeocyathids. In contrast, when one thinks of Paleozoic marine rocks, the first creatures to spring to mind are brachiopods, particularly the calcareous-shelled articulate brachiopods—the bivalved "lampshells" already briefly encountered. Present as modest (in body size and number of different species) members of the biota since the dawn of Cambrian times, the brachiopods suddenly exploded into a tremendous array of orders and families, shapes and sizes soon after the Ordovician began. It is virtually impossible to visit a post-Cambrian Paleozoic outcrop and avoid seeing a brachiopod. Indeed, their sheer physical presence in a way works against them: they are so easy to collect, so obvious, that many collectors automatically turn to the rare, hence exotic, fossils, turning up their noses at all but the most gorgeous brachiopod shells. In the Paleozoic, to find a clam is generally a source of joy, while in Mesozoic rocks, where clams are frequently abundant, the shoe is on the other foot and it is the brachiopods, by then greatly diminished in number after their Paleozoic heyday, that become the prizes. I vividly remember, in what was only my second experience hunting fossils in the field, rapidly becoming disenchanted with the three strikingly handsome species of Cretaceous oysters strewn all over a stream bank in southern New Jersey; I and my colleagues soon began to pay them no heed at all, mindlessly crunching them underfoot as we sought the rare, elusive single species of brachiopod in the fauna.

Brachiopods and clams have long been associated in the paleontological mind, even though they are not particularly closely related: brachiopods are so-called "lophophorates," while clams are bivalved, headless mollusks. Brachiopods derive their name from their fleshy tentacles (the "lophophore"); "brachiopod" means "arm-foot," an allusion to the lophophore. The lophophore fills most of the shell's interior, and is used for both respiration and straining minute food particles from seawater. Most brachiopods are attached in some fashion to the seafloor, or to other organisms, including floating patches of seaweed.

Yet we continue to associate clams and brachiopods in our minds, for the simple reason that both sport two shells composed of the mineral calcium carbonate ($CaCO_3$)—the basic mineral constituent of limestone. Not that it is terribly difficult to distinguish a clam from a brachiopod specimen: nearly all clams (check this the next time you order them in a restaurant) have almost mirror-image shells. The two sides are nearly identical, and are considered the *right* and *left* halves of the shell. (You tell left from right by holding the specimen so that the tiny curved tip that marks the very beginnings of growth on each valve is pointed away from you; to your left is the left valve). Oysters are among the few exceptions to this rule of left/right clam valve symmetry, as they lie on their left valve, and the right valve is reduced to form a much smaller covering flap.

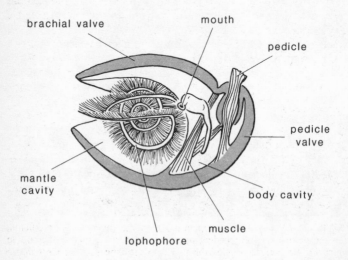

FIGURE 21 A diagram of brachiopod anatomy.

FIGURE 22 A slab containing many specimens of several different species of Ordovician brachiopods. Compare with the next illustration.

FIGURE 23 A slab of Mississippian brachiopods. The species are different from those shown in the preceding illustration, as are the higher-level groups to which they belong. Nonetheless this slab bears a striking resemblance to that in the preceding illustration—a strong hint that ecological conditions were quite similar when both were formed over 100 million years apart!

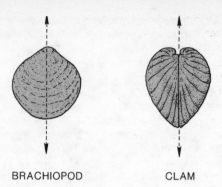

BRACHIOPOD   CLAM

FIGURE 24 A diagram illustrating the difference in symmetry between the shells of brachiopods and clams. For further explanation, see text.

Brachiopods are also symmetrical, but in a completely different sort of way. They have upper and lower valves: the upper, also called the brachial, valve, which supports the lophophore, and the pedicle valve, through which emerges the fleshy stalk that in many brachiopods attaches the organism to the seafloor. In brachiopods, the brachial valve is nearly always considerably smaller than the pedicle valve. You can perceive their symmetry, instead, by orienting a specimen with the brachial valve up, and with the beak (i.e., the point of earliest growth on the shell) away from you, and mentally slicing straight down through the shell: left and right sides will be nearly the mirror image of one another. In brachiopods, unlike clams, the plane of symmetry passes through both the brachial and the pedicle valve, dividing both into mirror-image halves.

There are, as you might well imagine given their remote relatedness, many other differences between clam and brachiopod shells. The internal musculature, which among other things is responsible for opening and closing the shells and supporting the soft, fleshy internal anatomy, is radically different in the two groups—as readily seen on fossil shells, as the muscles commonly leave indentations ("muscle scars") on the shells' interiors. And it is relatively rare to find a clam shell with the two valves articulated together in their normal "life" position; the reason is that the shell is opened not by muscles but by a sort of spring (the "ligament") tucked away at the upper edge, where the two valves are joined. Upon death, all the muscles (the "adductors") that hold the shell tightly closed relax, and the shell begins to gape. Clams expiring in refrigerated supermarket bins start that gape—a tip-off that you don't want to buy them. Brachiopods (which, though still alive in modern seas, are generally considered inedible) use separate sets of muscles to open and close their shell. Lacking a ligament to force the valves apart, in death the shells typically remain closely associated unless forced apart by scavengers or waves and currents.

The confusion between clams and brachiopods is by no means limited to the amateur who has yet to master the rules of symmetry and gain sufficient anatomical knowledge to tell brachiopods from clams in the field. Professionals still associate the two groups—because they both have two shells, and commonly are stuck to or simply lying on the seafloor, where they patiently, if rather mindlessly, strain nutrients from the seawater. Another reason for the association is that clams remain mostly a minor part of the Paleozoic faunas otherwise so dominated by brachiopods—and then the tables are turned, and from the Mesozoic up to the Recent there are far more clams than brachiopods. There has been an irresistible urge to see cause and effect here, to read a profound evolutionary message in the switch in relative abundance between these two great bivalved groups.

The issue is important because a persistent theme in this book is that extinction commonly opens the ecological space that sets the scene for what will happen next in life's evolutionary history. As at the Cambrian/Ordovician transition, we might expect the "invention" of wholly new groups, the disappearance of others, or the more subtle switches in relative abundance and ecological dominance of different forms of life. It remains, for example, a strong temptation to see the well-entrenched dinosaurs, occupying a wide variety of ecological niches, keeping the mammals down—a plausible explanation of the fact that mammals remained rat-like for the bulk of the Mesozoic. It was only after the dinosaurs had all gone that mammals underwent their own expansion and radiations.

So why not look at the clams and brachiopods this way? Paleontologists S. J. Gould and C. B. Calloway have recently taken a close look at the old story of supposed competition between clams and brachiopods. Rather than seeing clams as more advanced, hence competitively superior to the more lowly brachiopods, they believe that clams finally edged out brachiopods by sheer dumb luck: during the massive Permian extinction that terminated the Paleozoic, wiping out over 90 percent of all marine species, the clams were lucky. They managed to get a foothold as the shallow seas once again washed over the continental interiors and simply took over the vacant space as pioneers, thus asserting their claim in a manner that had been unavailable to them.

Yet we might go further and ask whether it is appropriate to compare clams and brachiopods at all. After all, most Paleozoic clams lay on the seafloor; their great post-Paleozoic expansion was not in this habitat, but rather seems to have involved an invasion of the sands and muds of the seafloor: for some reason not entirely clear, clams became burrowers to a far greater extent than they had ever been before. Clam paleontologist S. M. Stanley points to the development of improved siphons for the intake of clean water and the expulsion of stale, siphons protruding from the rear of the shell that enable a clam to remain buried indefinitely while still feeding and respiring, happy as a clam. Perhaps it was this key innovation of the siphons that triggered the radiation.

Insofar as competition for niche space is concerned, it is true that most clams and all brachiopods feed by filtration, but so, too, do a whole raft of other marine invertebrates. Thus there is no particular reason concerned with feeding that might lead us to suspect that clams and brachiopods were competitive. It is simply because brachiopods and clams have two calcareous shells that we tend to overdo the comparison between them.

The Great Ordovician Evolutionary Burst

Nevertheless, this sort of gross ecological comparison between groups can be instructive. When we make a list of creatures that either appeared for the first time or increased dramatically to round out the complexion of Paleozoic marine life in the Ordovician, it is immediately obvious that all such groups, representing a diverse array of genealogical backgrounds, are astonishingly uniform ecologically. All have shells composed of calcium carbonate. All are filter feeders, rooted or cemented to, or simply reclining on, the substrate, which is usually simply the seafloor itself. And the list of new or expanded groups is impressive: sponges, corals of several kinds (some of which, ironically enough, are probably really sponges), bryozoans of several orders (bryozoans, like brachiopods, are lophophorates). Among the echinoderms, cystoids, blastoids, and crinoids appear, while the clams and articulate brachiopods become important for the first time. In contrast, while the bottom-clinging, filter-feeding calcareous-shelled invertebrate organisms were expanding explosively, trilobites, snails (gastropod mollusks), cephalopods (particularly, in the Lower Paleozoic, the shelled nautiloids, of which the pearly nautilus is still extant), graptolites (extinct, colonial organisms) and a number of other groups failed to show any expansion—or diminution. What was going on, and why?

There is a pseudo-phenomenon in paleontology known as the "monographic burst." Those interested in general patterns in the history of life (arguably paleontology's greatest potential gift to evolutionary biology) tend to consult the technical literature, the monographs and papers detailing the history and classification of various groups of fossils, and simply count up numbers of species, genera, families, and so forth alive during any particular interval of geologic time. The *Treatise on Invertebrate Paleontology*, (now over twenty volumes long, with more on the way) is the handiest such source for invertebrates these days, but prior to its appearance (in the late 1950s), the compilers had to turn to the "primary" literature itself. Skeptics—mostly monograph writers who doubted the expertise of those who consulted their work—were quick to realize that one might obtain a very biased estimate of the true numbers of species or genera of a group simply because the fauna of some time periods were far more intensively studied and written about than others. The work of G. Arthur Cooper, the foremost student of brachiopods of the twentieth century, has been cited as a case in point. Brachiopods began in the Lower Cambrian and are very much alive today, hidden away in rocky overhangs, the recesses of reefs, and on the floor of the deep sea. Cooper has studied brachiopods across the entire spectrum of their geologic range. His best known and most extensive work, however, involves definite, restricted segments of geologic time: the Lower Middle Ordovician, the Devonian, the Permian, and, most recently, the Mesozoic and modern brachiopods. It is the Lower Middle Ordovician that arrests us here: could it be that the big explosion of sessile filter-feeding invertebrates in the Lower Middle Ordovician is a mere artifact of the prodigious efforts of G. A. Cooper and others in concentrating their efforts, for some reason, on that time interval rather than other periods of time? Or is something real going on in that period?

The answer is tricky. Something real did go on, and it is abundantly clear by now, particularly since far more than brachiopods were involved, that the reason for all those monographs is that so many fossils suddenly showed up in the rocks. The monographs reflect the diversity, rather than create it artificially. But when we look a bit harder at why the diversity jumps, we see that to some extent it is an artifact of sampling. For in the Lower Middle Ordovician the fossil record (certainly in North America and in Eurasia as well) all of a sudden gets markedly better. There are more rocks spread out over more extensive expanses and new habitats—extensive seas whose bottoms are floored with carbonate-rich muds; and there was the final development of widespread, full-scale continental seas that periodically flood a continent from side to side, warm tropical seas that were havens for a greatly expanded selection of multicelled animals.

Limestones are composed of the calcium carbonate mineral calcite (or, more rarely, the mineral aragonite, the less stable version of the same chemical compound). Precambrian limestones are known, but extensive study of the stratigraphic history of limestones reveals the vast bulk of them to be Ordovician or younger: the reason, in a way, is simple, though it poses a chicken-or-egg dilemma as far as the Ordovician proliferation is concerned. For most limestones are *biogenic* in origin—that is, the mineral content is derived from the shells of organisms. Shells occasionally accumulate as recognizable fossils. But most invertebrates make their greatest contribution to the rock record as minute granules of sediment that accumulate over the eons, eventually hardening to form the limestones and limy shales so common over both the low-lying interiors of continents and in the flanking mountain ranges of the continental margins.

Thus the carbonate blankets that begin to cover the interiors of continents are the result of a vast upsurge of carbonate-shelled creatures, not the cause of their population explosion. Yet an expansion of habitat allowed the proliferation in the first place. And it is difficult to resist thinking that once lime became a dominant constituent of the shallow inland seafloors, such a change in typical substrate would not have had further implications for the organisms crawling over, lying on, or burrowing down into those ancient seafloors.

Of Corals and Other Reef-Builders

Until quite recently, the received truth had it that corals did not appear until the Middle Ordovician inventive explosion. Recently, however, two species of simple horn (or "rugose") coral from the Middle Cambrian of Australia have been described. Even so, it is clear that corals really did not get going in force until the Ordovician—once again posing the old scale-of-nature problem. Corals are coelenterates, with conical bodies composed of two layers of simple tissues, and wholly lacking true organs of any kind. So how could it be, mused evolutionary-minded paleontologists, that animals as primitive as corals arose so late in time, or (after their eventual discovery in the Cambrian) expanded so late in their careers? As we saw with the Ediacara fauna, though, it is obvious that entirely soft-bodied coelenterates (sea anemones being a modern example) have long existed; occasionally some of them acquire the ability to secrete a protective calcium carbonate envelope and *voilà*—they begin to show up in the fossil record.

Modern corals apparently have little directly to do with their Paleozoic relatives—that is, from a direct ancestor-descendant point of view. The issue is still contentious, but after the Permian extinction, which took both the Rugosa and the Tabulata (now classed as sponges by some modern workers), modern corals (hexacorals, or, more formally, the Scleractinia) show up. Scleractinia include very

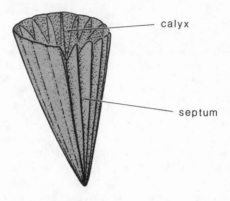

calyx

septum

FIGURE 25 Diagram of coral anatomy.

FIGURE 26 The Ordovician sponge *Brachiospongia*. While not overly common as large fossils, true sponges have been contributing to the fossil record since the Cambrian.

FIGURE 27 The Upper Devonian glass sponge *Hydnoceras tuberosum*. Sponges lack true tissues, yet have several different sorts of cell types, reflecting a division of labor. Sponges are classified as "Parazoa"—they fall just short of being true multicelled animals, or "Metazoa."

nearly all modern corals, and are close relatives of the sea anemones. The vertical walls that radiate around the simple cylindrically shaped internal cavities of these organisms, and which are known as "septa," simply increase the amount of surface area used for respiration and the absorption of food. These walls are arranged in six-fold geometric symmetry, but nothing like hexameral symmetry is to be found in the Paleozoic. The Rugosa, undoubtedly corals, are distinctly tetrameral—their symmetry comes in fours. The mineralogy of their skeletons, too, differs: the carbonate calyces of Paleozoic Rugosa were of calcite, while Scleractinia utilize the alternative form of calcium carbonate, aragonite, to fashion their chambers. It is possible, of course, to imagine a transition from one form to another over that missing interval of time at the Permian-Triassic boundary. Possible, but seemingly not likely, for the Rugosa had remained stable in these basic aspects of their anatomical organization from their Cambrian beginnings to their disappearance. And particularly because modern scleractinians have close naked relatives (sea anemones), it seems plausible that corals are not so much a "natural" evolutionary group as they are hardened, filter-feeding coelenterate polyps, an ecological invention that appeared at least twice in the history of life. When the Permian crisis hit, the rugosans went forever, and corals were reinvented from some group of their naked collateral kin.

Paleoastrophysics

Cornell University, nestled in the rolling Devonian hills near the southern end of Lake Cayuga in central New York, has had a long series of distinguished paleontologists on its faculty, dating well back into the nineteenth century. Indeed, as I write, the President of Cornell, Frank H. Rhodes, is a paleontologist—a source of some pride to the fraternity. And prominent among Cornell's faculty until his recent retirement was coral expert John Wells.

Wells has devoted the bulk of his career to the Mesozoic and Cenozoic scleractinians. Yet there he was in the midst of the Devonian in upstate New York, far removed from modern coralline seas or any ready source of fossilized scleractinians. It was inevitable that one day he would turn his attention to the varied, abundant and in some instances downright spectacular Devonian coral fauna of New York State. (A bed of shale about six feet thick crops out sporadically in an area of about 150 square miles in central New York. It is best seen on the eastern shore of Skaneateles Lake, and is absolutely jammed with nothing but the calyces of rugose corals. It is simply an incredible sight.) What Wells managed to do with some of those Devonian corals, though, was in itself pretty spectacular.

FIGURE 28   *Cystiphyllum*, a Devonian solitary rugose coral similar to those used by John Wells in his calculations of the length of the Devonian year.

One thing that Wells had learned from his studies on scleractinians is that they tend to lay down a thin, fresh band of calcium carbonate on the top of their calyces every day. These fine, small growth ridges tend to be grouped into monthly bands; and modern corals may also show an association of such monthly bands into yearly collections of twelve.

Figuring that rugose corals of the Devonian were probably similar in overall habits of growth to their modern scleractinian counterparts, Wells assembled a collection of particularly well-preserved specimens and started to count their growth ridges. What he came up with may at first appear a bit surprising: there seem to have been 400 days in the year during Devonian times! At least what appeared to be annual bands consistently contained an average of 400 growth ridges. Pursuing Wells's lead, British paleontologist Colin Scrutton examined other Devonian corals—this time specimens that appeared to show monthly bands. Here the average count was 30.6, which, when multiplied by 13, yields 397.8. Assuming that there has been no measurable change in the actual length of the year—that is to say, in the length of the Earth's journey around the sun—there were evidently 399 days a year in the Devonian Period, divided into thirteen months of thirty-and-a-half days apiece. How could this be?

Astronomers had long supposed that the rate of the Earth's spin on its axis—the motion that determines the length of our day—should have regularly diminished over time due to tidal friction, the drag effect of the gravitational interaction between the Earth and its moon. They calculated that the spin should have decreased by about 2 seconds per every hundred thousand years—or about 7,600 seconds, or a bit over 2 hours, in 380 million years. This rough estimate called for 22 hour-long days back in Middle Devonian times; multiplying 365 (omitting leap year) by 24 we get 8,760 as the approximate number of hours in our current year. Divide by 22, the rounded-off estimate of the length of the Middle Devonian day, and we get 398, the predicted number of days in the Middle Devonian year based on estimates from tidal friction.

Science lives to a very great extent by numbers. There is a tendency to disparage "theory" as mere guesswork. But solid theory is often expressed as the logical, mathematical consequences of a set of observable, measurable physical relationships—as with the friction set up by the relative motions of the Earth and its moon, the relationship which gives us our daily oceanic tides and produces a detectable bulge in the rigid crustal rocks of the Earth itself. Astronomers knew the Earth must have been slowing down in its rate of spin, and that it must have been doing so according to a smooth, mathematically regular and calculable rate. The numbers seemed secure.

But it is always nice to have some independent evidence from the real world to confirm our calculations. We seek to test ideas about the physical world by asking: what must be observably true if the theory is correct? Einstein's credibility went way up when Sir Arthur Eddington announced that the bending of light waves from remote stars as the light passed the sun (detected during a total eclipse) conformed to the theoretical expectations derived from Einstein's General Theory of Relativity. Einstein claimed not to be too taken with the results—his faith in the correctness of his theory was so strong. But being human, he was, of course, pleased. And the rest of the world, physicists and laymen alike, seized upon the experiment as the crucial bit of evidence needed to persuade them that the peculiarities of space-time were a reality. Einstein's equations describe the universe as it really is better than Newton's, and it took some empirical demonstration to convince the world that this was so. And Wells's and Scrutton's patient work on Middle Devonian corals is very much the same sort of affair: Wells saw that there was a possibility that the length of the Devonian year was recorded by corals living then, just as their skeletons record it now. It was, in retrospect, a perfectly obvious test of a prediction arising from physical theory. And it worked!

Reefs Through Time

When one thinks of corals, one thinks of reefs: massive brain corals and more delicately branched staghorns rising off the seafloor in a jumbled array, offering nooks and crannies for myriad invertebrates, and providing the basis for a complex ecosystem including the fish and swimming invertebrates that live just off the reef front. Reefs commonly grow up into the zone penetrated by the bases of waves, becoming topographic features that affect the development of the land surface as well as creating hazards to navigation. And ancient reefs have had a host of other, unexpected consequences: the dead Middle Silurian reefs ringing the Michigan Basin apparently so restricted the flow of fresh saline waters into the interior of the basin that what is now the entire southern peninsula of Michigan became a gigantic salt pan throughout much of the Upper Silurian. Still the major source of table and various rock salts, Michigan has salt deposits in excess of 1,600 feet thick near the very center of that ancient bowl.

Other reefs—such as some Devonian coral reefs buried deep below the surface in western Canada—have proved to be excellent traps for petroleum products. Liquids and gases trapped below the Earth's surface must be concentrated before they become economically feasible to extract. The rocks in the area must be sufficiently porous to accumulate the needed amount of oil; and the rocks must also be permeable—allowing liquids to creep along so they can be concentrated in one place. Reefs fill the bill very well, for all reefs since the invention of colonial organisms in the Cambrian consist of building blocks—usually, but not necessarily or exclusively, composed of corals—plus other creatures that serve as cementing agents holding the frame together.

Reefs are a great example of a biological "idea" that keeps being repeated even though the precise cast of characters has changed rather markedly over geologic time. All it takes is a colonial organism—anything that sits there and buds off new individuals which remain attached to the parent from which it sprang. We have already encountered the first colonists of the seafloor that met these requirements—the Cambrian archaeocyathids (formerly believed to be colonial sponges, but now considered more probably some form of algae). But true sponges make colonies as well, and have substantially contributed to reefs off and on since the Ordovician. Occasionally sponges were even the dominant element of reefs. In the Paleozoic, reef-formers also included massive colonial tabulate "corals" whose coelenterate affinities are likewise subject to considerable doubt (they seem much more like certain calcareous sponges recently discovered hidden away in the interstices of modern reefs). *Favosites*, perhaps more than any other Paleozoic "coral," most closely resembled modern colonial, reef-building corals. On the other hand, the lacy colonial tabulate *Halysites*, looking for all the world like Christmas ribbon candy, seems far too delicate to have served as the structural building block of a coral reef, yet appears to have filled that role admirably in many an Ordovician, Silurian, and Devonian reef setting.

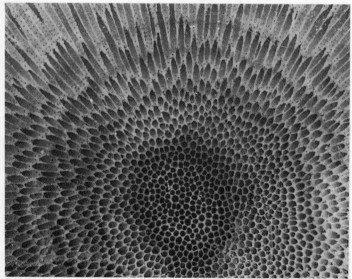

FIGURE 29 A polished surface of the Paleozoic tabulate coral *Favosites*. Each opening housed a separate coral polyp.

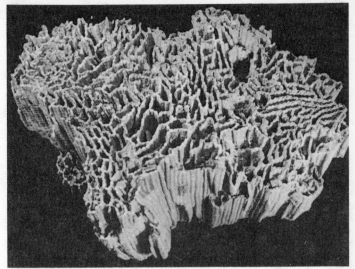

FIGURE 30   The ribbon tabulate coral *Halysites*.

Corals and Ancient Geography

Modern reefs are pretty much limited to a narrow band extending 30 degrees north and south of the equator: they are eminently tropical features. The limits of their distribution correlate strongly with the distributional limits of the so-called "hermatypic" corals—those that can support photosynthesizing algae in their soft tissues. Whatever the precise benefit in terms of internal energy budget the algae can offer their coralline hosts (the matter is still open to some dispute), one thing is abundantly clear: the hermatypic corals are the great, massive colony builders. Nor is this sort of algal symbiosis limited to the hermatypic corals: the giant clam *Tridacna*, a common element of Tertiary and modern coral-reef communities of the Indo-Pacific region, also houses algae in its fleshy tissues, near the gap between its two shells. Presumably, the metabolic boost afforded the clam by its single-celled algal invaders is what helps *Tridacna* attain its monstrous size, familiar to many people from museums, shell shops, and *New Yorker* cartoons. And relatively recently, biologist John Lee has discovered algae living in the tissues of giant, single-celled foraminiferans, which are a form of shelled amoebae. From a paleontological point of view, Lee's discovery is extremely interesting, as there are a number of extinct groups of truly giant foraminiferans, and Lee has seen evidence that they, too, as one might well expect, played host to photosynthesizing guests in their protoplasm. The pyramids at Gizeh, just outside Cairo, are made of Eocene limestone quarried from the Moqqatam Hills on the other (east) side of the Nile, and are chock full of *Nummulites* (meaning, literally, "money-stone") and related single-celled foraminiferans, some as much as two inches in diameter. In the Paleozoic, toward its end, lived the fusulinids, cigar-shaped protozoans wound up like jelly-rolls. Some of these attained lengths of up to four inches or so—amazing for a single cell! Lee thinks these were undoubtedly as "hermatypic" as their coralline counterparts.

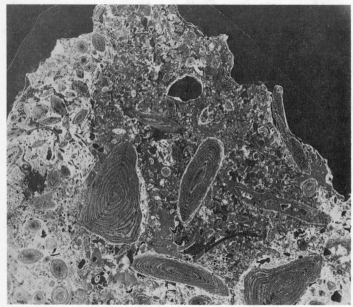

FIGURE 31   A polished piece of Eocene limestone from Egypt, exposing cross-sections of the gigantic shelled amoeban *Nummulites*.

Again, just as John Wells assumed that ancient corals grew in much the same way as their modern counterparts, it seems reasonable to assume that the massive reef-building colonial corals of the Paleozoic were also hermatypic, and thus probably also confined to a belt more or less parallel with the equator in those days. And it is no perfectly straightforward matter to identify where the equator *was* back then, because although we assume that the Earth has retained its same basic orientation since the Paleozoic, we now know that the continents have not. Thus the equator must have run through different parts of the continents and ocean basins than we see it running through today.

Conveniently enough, there are two independent general ways to determine where the equator *was*, in terms of present-day continents and oceans at any remote time back in the Earth's past. One technique is purely physical, and dates from some of the final, clinching geophysical evidence that convinced everyone back in the 1960s that the continents have indeed been moving about *vis à vis* one another. It seems that particles of some iron-bearing minerals are magnetic: lodestone, of course (the mineral magnetite, Fe_3O_4) but also to some lesser degree the red iron oxide hematite (Fe_2O_3), which, when concentrated sufficiently, forms the principal ore for iron. Hematite is disseminated in lesser quantities in many sedimentary deposits—such as the red beds of the Upper Devonian of the northeastern United States (and the "Old Red Sandstone" of England and Scotland). It is very abundant, as well, in the reddish brown sediments of New Jersey and the Connecticut Valley, sediments of the uppermost Triassic and Lower Jurassic ages that have produced the remains of dinosaurs and other Mesozoic creatures. The weak magnetic orientation of the iron minerals in rocks such as these reflects the position of the Earth's magnetic poles at the time when the sedimentary particles were deposited; the orientation can be measured with sensitive magnetometers. According to this method, the Devonian north pole proved to be located in the North Pacific, indicating that something was clearly amiss in the traditional pre-Continental Drift thinking. (When the pole proved to be in two different places, depending on whether one was measuring its orientation from eastern North America or Great Britain, there was hardly any choice but to concede that the continents had been moving about—assuming that there has always been only one north magnetic pole at any one time.) And if we can measure the ancient pole positions, plotting them on a map of the Earth's surface, it is no great trick to move 90 degrees up and down from the magnetic poles to find that central line we call the "equator."

Still, however, as with Wells's corals, it is nice to establish some evidence that corroborates the calculations and fleshes out our picture of ancient geography. And that is done by examining past distributions of organisms whose modern counterparts are known to have characteristic distributions paralleling such geographic features as the equator. Maps of ancient coral-reef distributions agree nicely with the reconstructed locations of the continents in bygone epochs. The tremendous development of coral reefs in places like Michigan in the Middle Paleozoic was possible because the equator at that time ran across North America from northern Mexico or southern California in a "north east" direction (i.e., from the perspective of modern geography) up toward the northeastern region of the continent.

By the same token, the south pole in Lower Devonian times seems to have been but a few hundred kilometers north of Cape Town. At that time, what we now know as "Africa," "South America," "Antarctica," "Australia," "India," plus their associated island regions was all one huge continental mass, the famous "Gondwana." (Gondwana means "Land of the Gonds"—a people living in modern India; "Gondwanaland" thus ranks in there with the "Rio Grande River," "Gobi Desert," and other such redundant place names.) Shallow seaways covered parts of Gondwana during Upper Silurian and Lower Devonian times—seaways absolutely teeming with marine life. Yet a comparison between Lower Devonian rocks and fossils of the Appalachians, say, or the region just south of Prague in Czechoslovakia, and those of Gondwana of exactly the same age reveals a great deal of difference between them. One might expect that North America and Europe were closely associated, so we are not surprised that the faunas from those regions resemble each other a great deal more than either looks like Gondwana faunas. Fair enough. But there is more to it than that. For example, the great bulk of European and North American sedimentary rocks of the Upper Silurian and Lower Devonian are limestones—indicative of rather warm conditions (as corroborated by present-day experience: most limy sediments today are found on shallow tropical and subtropical seafloors). In contrast, hardly a trace of limestone is to be found in the same aged strata of Gondwana. And while different rock types obviously reflect different environmental conditions on those ancient sea bottoms, this lack of evidence is absolutely consistent with the paleomagnetic evidence that only a few non-reef-building corals have ever shown up in the thousands of exquisite fossils recovered from those southern hemispheric Siluro-Devonian rocks. And those groups that are present—notably, from my point of view, a tremendous radiation of trilobites—have many species that are extra large and quite spiny, typical of colder water environments. Once again, we see physical and biological evidence converging to solidify our interpretation of the world in remote times.

Coral reefs offer one additional very valuable lesson about life's history. All organisms are "adapted" to their environments. And most of the adaptations, reflected in their anatomy, behavior, and physiology, are concerned with ways and means of surviving, of "making a living" extracting energy from the surrounding biotic and physical environment—and of withstanding the physical and biological threats to their existence. (Some other adaptations are concerned purely with life's other grand theme: reproduction.) Often there may be one or at most only a few ultimate "best ways" open to organisms to perform a certain function. Because of the limitations of organic design (no biological system has come up with wheels, for example—and more than likely none ever will) and because the problems of existence are nearly the same for many organisms only remotely related to one another, the phenomenon of "convergent evolution" is strikingly common, and has attracted the attention of evolutionary biologists for many years.

The basic design of a single coral organism is simplicity itself: as we have seen, it is a conical tube, lined with a double-tissue layer that performs all the essential functions of feeding, excretion, respiration, and reproduction. Many other creatures, though, are also sedentary, gleaning a living by straining particles from seawater. Thus it is no real surprise, but nonetheless remarkable, that a number of totally unrelated groups have adopted essentially a "coralline" habitat, particularly when living in reef-like settings which may or may not also incude the presence of true corals. In the Permian, toward the very end of the Paleozoic, reefs were heavily developed in many regions of the world. An entire group of brachiopods developed extraordinarily deep, conical pedicle valves—and resemble corals so much that it takes a careful look inside to tell the difference. In the Cretaceous, toward the close of the Mesozoic, clams turned the trick: the rudistids, greatly enlarging their right valves, adopted a coral habit—but in a giant way. Forming genuine reefs (as, for example, seen in Cretaceous deposits of Cuba), these huge clams almost certainly developed the habit of modern giant clams (*Tridacna*) of nurturing colonies of photosynthesizing algae to enable them to attain their prodigious sizes. Time after time, in group after group, the coral-habit, with many of its ecological correlatives, such as algal symbiosis and a reef-forming habitat, appeared: evidently a good ecological/evolutionary idea èxploitable by many a group faced with similar exigencies of life in the sea.

Life in the Mid-Paleozoic

You get a sinking feeling standing in one of the gigantic limestone quarry operations scattered around the midwestern United States. On the one hand, you realize that without the very act of quarrying, the terrain would have remained gently rolling forest or farmland—with no hope whatsoever of sampling any of the millions of fossils that lay below. Yet a paleontologist feels more than a little wistfulness at seeing ton after ton of rock hauled off to the crusher, the fossils in them destined soon to become powdery morsels of concrete.

Such lugubrious thoughts tend to be forgotten as you begin to take advantage of the collecting marvels that present themselves in the spoil heaps and fresh walls all over these quarries. It is not uncommon that over 200 different species of invertebrate marine animals will be present in the mid-Paleozoic—some in utter profusion, others so rare that careful searching for years may produce few if any specimens.

Some of these midwestern quarries are huge. United States Steel runs the largest limestone quarrying operation in the world—in Silurian and Devonian rocks just outside Rogers City, along the eastern shore of Michigan's southern peninsula. More human-sized are the quarries in Petoskey and Alpena, Michigan, in Sylvania, Ohio, and in Waldron, Indiana. Nor are such quarries strictly limited to the Midwest: Becraft Mountain at Hudson, New York is yet another typical operation, where limestone is systematically removed for the manufacture of cement. None of these quarries is in any sense "small": you can easily become disoriented while trudging over and around vast stretches of spoil heaps (where discarded rocks often repay the careful scrutiny of the assiduous fossil collector) as the heat waves shimmer on a sweltering summer's day. I once spent well over an hour of precious collecting time simply trying to relocate my equipment.

Sitting down on a flat, exposed "bedding plane" (the hardened remains of an ancient seafloor), prising out specimen after specimen, you begin to get a feel for the fabric of community life in mid-Paleozoic seas. Four or five species of trilobite are commonly present, though seldom represented in overwhelming abundance. Brachiopods, on the other hand, are usually extremely abundant: delicately winged spiriferids and their close kin may contribute up to twenty or thirty different sorts of brachiopods collectable in a single quarry, though ten is a more likely number of brachiopod species living in close association, the remnants of past ecological communities.

Joining brachiopods in the filter-feeding bottom-dwelling cadre are crinoids (sea lilies), rooted to the bottom by stalks. Though crinoids are even more spectacularly prolific in the later Paleozoic, nonetheless each bottom-dwelling community seems to have supported on the order of five or perhaps ten different crinoid species, suspended at various altitudes off the bottom—presumably to divide up the available resources. Confusingly similar blastoids, whose tulip-bud-like bodies sprouted five rows of "armlets" (as opposed to the five feathery arms of true crinoids), likewise stood fast on the bottom, suspended in the bottom currents by long, flexible stems. And as with the crinoids, it was not until the Mississippian (Lower Carboniferous, to all but American geologists) that the blastoids really became abundant and diverse.

Corals, as we have seen, were quite common in these ancient communities. However, corals tend to come in crowds: though not unknown, simple stray specimens of, say, solitary rugose corals are still rather uncommonly encountered. Usually, when you have found one coral, you have found a whole raft. Sponges are not overly conspicuous—at least as body fossils; however, their microscopic skeletal pieces ("spicules") are fairly common in muddy residues of dissolved rock samples, proving that sponges indeed were far more common elements of Paleozoic communities than we might otherwise suspect.

But even here we have not exhausted the list of bottom-dwelling filter-feeders. The bryozoans, for example, those tiny colonial-living relatives of brachiopods, were very much in evidence in all of these communities. Clams, too, were there, but as we have seen, in far less profusion than the superficially similar brachiopods. But clams point out another aspect of mid-Paleozoic ecology—a point well worth noting in general, as it conforms so closely to the world as we see it today. Recall that T. H. Huxley warned us a century ago that there are two sorts of mistakes we might be making when using fossils to correlate rocks: we might claim that similarity in fossil content implies near equivalence in age—only to have misjudged the longevity of some assemblage. We would then have erroneously correlated two samples of widely different age. And we might be impressed by the dissimilarity of fossil faunas, concluding that they differed markedly in age, only to have failed to see we had two collections of the same age. Paleozoic clams illustrate the latter principle nicely: though by no means restricted entirely to the shallow waters just offshore, Paleozoic clams nonetheless are far more diverse, abundant, and easily observed in these muddy, shallow water environments hugging the margins of ancient inland seas than in deeper waters. Paleontologists Peter Bretsky, A. M. Ziegler and A. J. Boucot have all carefully examined the distributions of these old sea-bottom communities, finding upwards of five or six communities roughly parallel to shore, extending from near-shore situations dominated by clams, certain types of snails (the primitive "bellerophonts"), and trilobites to the quieter, deeper waters offshore, dominated by various combinations of brachiopods. Since they all lived at the same time, and all looked so different from one another, it is tempting indeed to assign them to different slots of the geological time scale.

Carnivores of Paleozoic Seas

I have said little so far of the carnivores that lurked in the Paleozoic seas. Paleontologists have speculated from time to time that trilobites, which roamed over the bottom and perhaps occasionally even swam a bit, actively fed on other creatures the way advanced crustaceans (such as some crabs and lobsters) do today. But trilobites lacked anything resembling true jaws, and it seems far more likely that trilobites relied on roiling up the bottom and straining particles of decaying flesh, minute crustaceans such as ostracods, and the omnipresent single-celled protozoans, algae, and bacteria.

But there were true carnivores back then—and here, finally, we get to the vertebrates. "Fish" first appear in the Middle Ordovician, as scrappy bony plates, scattered about in the rock—hardly prepossessing fish fossils at all. I place the word "fish" in quotation marks advisedly, as that four-letter word covers a vast range of creatures—well over half the kinds of vertebrates, fossil and living, extinct and alive today. Today's "fish" include (1) two types of jaw-less fish, (2) sharks and ratfish, and (3) bony fish, a group itself divided into: (a) the "living fossil" *Latimeria*, the only surviving coelacanth, (b) lungfish, likewise considered "living fossils" and living today in a remnant Gondwana distribution, and (c) "true bony fish"—salmon, swordfish, guppies, pickerel, bass, and many more. Within that latter group, the vast majority of all living "fish" species, there are many divisions, including a few surviving remnants of groups much more common in the fossil record of long ago.

FIGURE 32 Head and shoulder armor of the Devonian arthrodire *Dunkleosteus.*

In simpler times the Devonian Period was often called "The Age of Fishes" in unabashed acknowledgment of our seemingly endless preference for vertebrate animals. Indeed, the Paleozoic gives us a glimpse of a far more richly diverse fish fauna than the list of today's fishes. All modern fish are represented in the record—some (such as lungfish and coelacanths) in far greater abundance than today. But there are also groups in the Paleozoic that have simply disappeared—fish like the fearsome arthrodires which in some cases grew to lengths in excess of twenty or thirty feet, and were equipped with vicious shearing jaws. Arthrodires had a head encased in bony armor, attached by a sort of swivel arrangement to a shoulder armor apparatus. No one knows what other fish group comprised the arthrodire's closest relative: they simply show up in the Lower Devonian, evolve like mad, and become extinct by the end of the Devonian. In the Cleveland Shale, near the very last gasp of Devonian times, huge nodules scattered throughout the fine-grained (and otherwise nearly unfossiliferous) black shale have produced exquisite, three-dimensionally preserved specimens of these enigmatic, impressive fish.

Jawed fishes first show up in the Upper Silurian in the form of acanthodians, another group of uncertain affinities that managed to live to the end of the Paleozoic. Prior to their appearance, two sorts of jawless fishes (related to modern lampreys and hagfishes, the two sorts of skeleton-less jawless fishes alive today) made their presence known, but these were presumably not actively engaged in a carnivorous mode of existence.

The Devonian also saw the emergence of true bony fishes—in fact, all major sorts appear in the Lower and Middle Devonian (along with the last major fish group—the sharks). I have referred to both coelacanths and lungfishes as "living fossils"—and their histories, parallel in many ways, reveal much of what is meant by that rather overly romantic term.

Coelacanths and the Phenomenon of "Living Fossils"

Science in large measure lives up to its reputation for having a penchant for precision. Certainly loose usage in scientific terms does nothing to alleviate confusion or sharpen our appreciation of how nature is organized, how it functions, or what its history was like. "Living fossils" means many things to many people, and to some, such as the late Thomas J. M. Schopf of the University of Chicago, the term stood for no real phenomenon at all. Schopf had a point: recent work at the molecular laboratory bench has only confirmed the suspicion that portions of the DNA sequences in most species are pretty much in a state of flux: the old expectation that change is inevitable given the mere passage of sufficiently long chunks of time seems corroborated by the common observation of the rapid overhauling of many DNA sequences as observed in modern genetics laboratories.

Yet there are groups of living organisms that seem to have changed very little in their physical appearance from the anatomical configuration of their ancient forebears. Living fossils are not merely members of ancient groups—after all, we are all descended from the most ancient of all ancestors, those primordial cells of 3.5 billion years ago. Sponges would all have to be classed as "living fossils" if membership in an ancient group were the criterion—and the term "living fossil" would indeed be an empty phrase. A closer look at some typical and oft-cited examples, such as coelacanths and lungfishes, helps to bring the concept a bit more into focus, and to highlight some recent themes in the paleontological side of evolutionary theory. For "living fossils" seems to embody some interesting notions after all.

The first recorded catch of a live coelacanth was off the coast of South Africa in 1938; a fisherman dredged up a strange-looking, coarse-scaled fish about four feet in length. Suspecting he had something unusual, he kept the fish. Eventually it was inspected by a Miss Courtenay-Latimer, who recognized the fish from some photographs she had once seen in a book—photographs, no less, of fossil fish. These fish have a distinctively forked tail, with a fleshy lobe jutting out between the forks as a sort of terminal addendum. This characteristic is a dead giveaway that the fish is a coelacanth—a handy fact learned by all students who take up the study of fossil fish. Sure enough, this central fleshy terminus was present on the tail of the fish inspected by Latimer. Since then, the number of specimens has grown to several hundreds—though, curiously, no more have been caught in the waters off the southern African coast: without exception all of the subsequent specimens have been caught off the Comoro Islands, just to the north of Madagascar.

To this day, the youngest known fossil coelacanth is Cretaceous in age; at a length of around two feet, Macropoma is less than half the size of its living counterpart. Yet apart from its size, it looks rather like Latimeria—the modern coelacanth named for Miss Latimer. There are many other examples in which the youngest specimen known from the fossil record is actually far removed from modern times. Neotrigonia, a clam discovered in the nineteenth century off the coast of Australia, is the sole survivor of the Trigoniidae, a family otherwise known only from the fossil record. For years, the youngest

FIGURE 33   A living fossil, the coelacanth *Latimeria chalumnae.*

member of the Trigoniidae was Cretaceous in age. Further work filled in the record somewhat, as later in the nineteenth century specimens were recovered from Tertiary deposits in Australia. But so far, we still have no Tertiary coelacanths; presumably, shallow-dwelling coelacanths became extinct at the end of the Mesozoic (in the major extinction that dealt a death blow to so many groups of organisms). Only the deep-dwelling coelacanths escaped, and we have already seen the poverty of our available record of life in the true deep sea, the oceanic basins that are continuously recycled and rarely leave a fossil record accessible on land. Thus we can expect never to find the remains of Tertiary coelacanths even though we know they were present during that period.

Both coelacanths and lungfishes got started with a bang: from their earliest appearance in the Middle Devonian, both groups experienced a meteoric rise, becoming very diverse and reaching an acme in the Upper Devonian. The Gaspé Peninsula, which features near its very end Percé rock, with its millions of Lower Devonian invertebrates, is also home to several of the world's best known fossil fish localities—especially along the shore of Scaumenac Bay on the southern coast of the peninsula. Modern lungfish seem a strange lot, accustomed as they are to holing up for long periods as the seasonal rivers in which they live periodically dry up. These lungfish are long, skinny affairs, with little in the way of prominent fins. In the Devonian, in contrast, lungfishes came in many different shapes and sizes, including the prepossessing *Fleurantia,* my personal favorite, which was designed for all the world exactly like a rapacious mackerel, streamlined for a life of fast swimming and highly active predation.

But after the Devonian, both lungfishes and coelacanths seem to have "settled down" to a narrower range of anatomical styles. Indeed, very little in the way of change occurred after the initial bursts of anatomical diversification. In fact, the more unusual genera failed to survive the Devonian; it is as if the specialized adaptive types were weeded out. In any case, after the Devonian, there was little diversity in either the lungfish or the coelacanth lineages.

Thus, living fossils are simply species alive today that display relatively close anatomical similarity to ancestral species living way back near their group's inception. They tend to belong to groups that, at least for the vast bulk of the group's history, have not produced very many anatomically varied species. Except for brief flurries of diversification typically right at the beginning of the group's history, living fossil lineages seldom have more than a few basically similar species in existence at any one time.

There have been a number of theories advanced to explain the seeming incongruities posed by these "living fossils." Some paleontologists have been content merely to shrug the problem away, as if to say, "Well, some groups evolve quickly, the vast majority exhibit a moderate rate of evolutionary change, while a very few others have simply inherited the low end of the spectrum of rates of change." Such an attitude hardly *explains* why some groups evolve more quickly than certain others. Paleontologists have sought the explanation in genetic material: perhaps some groups, such as lungfishes and coelacanths, simply lack the requisite genetic variation that would allow them to escape the straightjacket of their ancient anatomical design. Still others have speculated that such resistance to change merely reflects a constancy of environment: natural selection simply keeps animals and plants looking the same as long as their environments remain recognizably constant. This last notion is, to my mind, more on the right track—yet it doesn't explain why still other groups *do* change even though they share the same supposedly constant environment.

Living fossils are something of an embarrassment to the expectation that evolutionary change is inevitable as time goes by. Paleontologist George Gaylord Simpson, who until his recent death was the earliest and most influential exponent of using modern biological approaches to paleontology, pondered over the discrepancy between apparent rates of evolution in clams versus mammals. Mammals seem to evolve perhaps ten times as fast as clams—as measured, for example, by the longevities and rates of appearance of genera. Mammals simply exhibit a higher turnover rate of genera than clams. Why? Some paleontologists have denied the phenomenon is real; T. J. M. Schopf and some colleagues, for example, speculated that since clams are far less anatomically complex than mammals, it is simply easier to record evolutionary change in mammals than in clams. But comparing evolutionary rates in groups of equal complexity—clams with clams, mammals with mammals, for example—still reveals a wide spectrum of rates. In general, clams probably evolve more slowly than mammals because marine creatures on the whole evolve more slowly than terrestrial organisms—a reflection, basically, of the more complexly heterogeneous environment on land than in the sea. But why do some mammals evolve more slowly than others? Why is *Latimeria*, for example, the endpoint of an evolutionary lineage that has remained essentially static since the Upper Devonian—for some 460 million years?

There seem to be some additional ingredients commonly present in most outstanding examples of this phenomenon of "arrested evolution." It has been noticed, for example, that the most primitive-looking modern species of any group are almost invariably the most "generalized" ecologically—their behavior and physiology, their very way of making a living, is typically more diverse than their more "advanced," anatomically more specialized close kin. They eat a greater range of foods and can tolerate a broader range of temperatures, pressures, and salinities (if, of course, they live beneath the waves). They tend, in short, to be jacks-of-all-trades. The standard assumption is that primitive means "lack of change," and the mystery of *why* they are resistant to change remains. But recently we have been thinking that maybe it is the other way around: being a jack-of-all-trades can be useful, particularly in variable and somewhat unpredictable environments. Perhaps there is some link, where the advantages of not becoming specialized actually increase survival probability and act as a sort of "brake" on evolutionary change. It is advantageous to specialize, to become proficient at making a living in a particular way—but for long-term survival, those who do not put all their evolutionary eggs in a single basket seem to have somewhat better odds.

And sure enough, in case after case, organisms from groups with many species are highly specialized, "advanced" over their closest relatives, which are often members of groups with far fewer species. Paleontologist Michael Novacek of the American Museum of Natural History recently surveyed the elephant shrews (an obscure group of small, insectivorous African mammals) and came up with that pattern. The flocks of cichlid fishes inhabiting the East African lakes provide another classic example. There is a growing conviction that ecological generalists evolve less quickly than ecological specialists because they tend to be far-flung in their occurrence and catholic in their tastes. They are thus better able to ward off extinction. But even more critically, from the standpoint of their

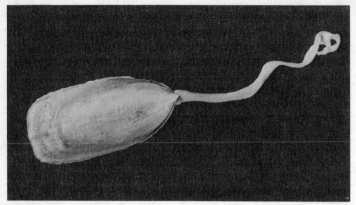

FIGURE 34   The inarticulate brachiopod *Lingula*. Note the long pedicle, used by these creatures to attach themselves to the substrate. Compare this modern specimen with the next illustration.

FIGURE 35   An Ordovician *Lingula*. Comparison of the complex anatomy inside fossil and recent shells reveals that these organisms have changed but little over the past four or five hundred million years.

resistance to change, such generalists appear very difficult to disrupt—to fragment into two or more daughter species as a response to environmental vicissitudes. And as I myself have long been arguing, together with Stephen Jay Gould and a number of others, most evolutionary change seems actually to occur in conjunction with this process of species fragmentation—"speciation." The condition of being an ecological generalist automatically decreases the rate of speciation in a lineage; and low rates of speciation ensure low rates of anatomical change.

FIGURE 36 The Cretaceous bivalve *Trigonia*, whose close relative *Neotrigonia* still lives in coastal Australian waters.

Other Carnivores

But fish were hardly the only carnivores back then: for one thing, the earliest jawed fishes, the acanthodians, do not show up in the fossil record until the Upper Silurian—and it is hard to imagine a marine world without carnivores for some 150 million years before jawed fishes appeared.

We get part of the answer to the question Where were the early carnivores? when we remind ourselves of one glaring source of bias in our knowledge of what was going on in those ancient times. Many creatures then, as now, simply lacked the sorts of hard skeletal parts that lead readily to fossilization. Many "worms" (there are several dozen animal phyla that qualify for the general sobriquet "worm") are actually carnivorous. But their soft squishy bodies are rapidly consumed and decayed. Worms, as a rule, don't stand much of a chance of preservation—though they occasionally show up as paper-thin smears on the partings of thin-bedded dark shales. We sometimes see microfossils of their jaws—minute yet effective tools for grasping and cutting prey.

Arthropods, though, are also commonly predators, and they have a hard, mineralized outer skeleton which is quite readily fossilized. Even if, as I firmly believe, trilobites themselves were not among the macro-carnivores, there were some truly large crustacean and horse-shoe-crab-like arthropods lurking among the seaweeds on Cambrian seabottoms. *Anomalocaris*, whose segmented form was for years interpreted as the body of some otherwise enigmatic arthropod, is now known to be simply fragments of the legs of a carnivorous arthropod that grew to several feet in length. Modern horseshoe crabs—another excellent example of "living fossils"—are the last surviving remnant of a large community of spider relatives inhabiting the briny deep. Most famous among them were the Paleozoic eurypterids—so-called sea scorpions. Specimens of *Pterygotus* from the same Cleveland Shale that produces giant arthrodire fish may have come from animals as long as fifteen feet. Looking for all the world like giant scorpions (the main difference is that scorpions have lungs for life on land, while eurypterids had gills), *Pterygotus* was equipped with a wicked looking set of pincers surely as deadly as the claws of a large Maine lobster.

FIGURE 37 An unusual "belly" (ventral) shot of a Silurian eurypterid from New York State.

But eurypterids usually, if not invariably, lived in brackish water lagoons, off the beaten track of Paleozoic marine life. Indeed, eurypterids, horseshoe crabs, and their ilk are important for our consideration of the animal invasion of the land, and so, perhaps oddly, we will meet them again in the next chapter. There must have been some other group of active, swimming carnivores back then. And there was: the cephalopod mollusks.

Modern cephalopods ("head foots") are highly advanced carnivores. We know them primarily as octopi and squid; but there is one other member of the clan, once again the last vestige of a once mighty and vast array of diverse creatures. Squid and octopi, as might be imagined, are poor candidates for preservation—though rare specimens of squids have been encountered even in the Paleozoic, particularly from unusual deposits such as the Essex fauna at Mazon Creek in Illinois, in which soft-bodied fossils are, amazingly enough, rather common. Some squids, though, had hard internal skeletons instead of the translucent "pen" that we find in modern squids. Beginning in the Upper Paleozoic and proceeding right through to the uppermost Cretaceous, belemnites—cigar-shaped hard spikes of calcium carbonate—are fairly common fossils, especially in the Middle and Upper Mesozoic. Thought by medieval Europeans to be petrified lightning bolts, these hard brown spiky objects have very few remaining features to remind one of squids—though squids they were, as some particularly well-preserved specimens attest.

It was the primitive, vestigial division of an otherwise extinct branch of the cephalopods that did so much of the carnivorous marauding in Paleozoic and Mesozoic seas. The modern pearly nautilus, represented by several living species of the Indo-Pacific, is a magnificent creature, memorialized in Oliver Wendell Holmes's poem "The Chambered Nautilus." The beautifully coiled shell of the most common species, *Nautilus pompilius*, adorns many a living room shelf to this day, as it has since the earliest days of trading with the Orient. Sliced in half, the shell gracefully exposes the internal partitions that completely close off its interior. The animal itself (that is, its head, tentacles, guts, and gonads) is restricted to the outermost "living chamber," pressed up against the last partition, the "septum." Periodically, as the animal enlarges the shell by adding to its growing outer margin, the body lets go of the last chamber, pulls itself forward a bit, and then lays down a new partition. The abandoned chambers are filled with gases; pressures can be regulated through a fleshy tube, the siphuncle, which is the last remaining means of communication between the living animal and its early, now emptied, chambers. The whole shell acts as a bouyancy device—and the nautilus can regulate its depth simply by varying the gaseous content of those chambers. Liquids partially fill several chambers to aid in stabilization.

Rather rare and restricted now to certain areas, nautiloids were all over the place in days of yore. We prize the living ones so highly not merely because of their beauty; without these particular living fossils, we would be very hard put indeed to understand the myriad

FIGURE 38 The last of the externally shelled cephalopods: *Nautilus pompilius.*

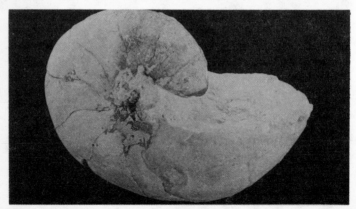

FIGURE 39 A Cretaceous nautiloid. The wavy lines on the sides of the shell reflect the rather simple suture patterns of all nautiloids.

fossil nautiloids—and their offshoot relatives the ammonites, which became one of the most important groups of animals ever to leave a mark on the history of life.

Although there were a few nautilus-like, small-shelled creatures in the Upper Cambrian, the nautiloid clan really seems to have exploded in the Early Ordovician—not the Early Middle Ordovician, when all those filter-feeding organisms really took off. In Lower Ordovician beds near Fort Cassin, Vermont, for example, a remarkable array of different nautiloid forms has been recovered. Again we find the familiar story: the greatest array of anatomically diverse species occurred during the earliest days of this group. Indeed, in a recent classification, the word "nautiloid" proper was reserved for a group which included modern coiled species as well as some straight-shelled and variously curved forms that date back to the Upper Cambrian. Yet, given the same high rank as true "nautiloids" (i.e., as a subclass of the Class Cephalopoda of the Phylum Mollusca) were two other groups that had previously been known simply as "nautiloids:" the actinoceroids and the endoceroids.

The funny thing about the array of early nautiloids is that they all look pretty much the same—externally, that is. They are all simple

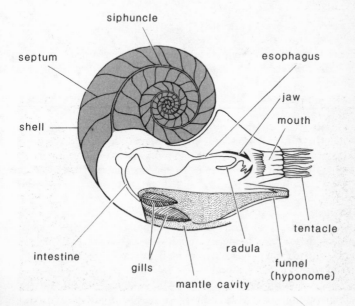

FIGURE 40   Diagram of cephalopod anatomy.

FIGURE 41 *Orthoceras*, a straight-shelled Ordovician nautiloid. The path of the siphuncle can be plainly seen running down the middle of this abraded specimen.

expanding, generally uncoiled (i.e., straight) shells; the differences between all of them are entirely internal. To get at their internal anatomy requires slowly cutting through the rock-filled shells, and polishing the slabbed surfaces. This reveals an astonishing variety of ways and means of utilizing the buoyancy potential of those vacated chambers common to all these creatures. All of them deposited layers of calcium carbonate; some had deposits only within the siphuncular tube, others only in the vacated chambers themselves. Still others had it both ways. And the form of those deposits varies. The layers lining the chambers imply the presence of a thin organic tissue throughout life—unlike in modern nautili. Nor was the regulation of gases in what remained empty in those old chambers ignored: in what is one of my favorite bits of ancient anatomical engineering, careful acid etching and polishing of some actinoceroids reveal a complex set of thin tubes for gas diffusion running between the siphuncle (itself lined with calcium deposits, thus not able to diffuse gases as in modern nautili) and the empty chambers.

Once again we see a great potpourri of anatomical styles, although all amounting to roughly the same thing—a variety of solutions to some common problem (in this case, matters of buoyancy: depth control and orientation of the shell). That only one design survived the Ordovician may not mean that the others were inferior or failed to do the job. The one that survived may simply have belonged to the luckiest lineage.

Some of those Ordovician "nautiloids" (in the broad sense) grew to lengths of thirty feet or more. Such huge beasts are generally depicted lying on the seafloor, sedentary giants waiting in ambush for their prey. Octopi, giant or not, hunt this way—though it is difficult to imagine how such huge shelled beasts as the nautiloids could possibly camouflage themselves from their would-be prey. In any case, it is safe to assume that the vast majority of these nautiloids swam in just the manner as their modern counterparts—by jet propulsion. Squids, octopi, and nautili all take water into a sack-like space (the mantle cavity, which also houses gills and sense organs); they have the capacity to close off the cavity, leaving only a fleshy tube, the "hyponome," from which they can expel the water with great force. Squids normally swim backwards—their streamlined "tails" leading the way, their heads with their bright and very good eyes facing the rear, their long tentacles flowing behind. But when they spot something to eat, they simply point that hyponome the other way and off they go, forward, tentacles outstretched to grab the fish, crab, or whatever other delicacy has caught their attention. In alarm, whoosh, off they shoot backwards, expelling a cloud of ink from their mantle cavities as they beat their retreat. Traces of ink from Paleozoic cephalopods have recently been reported. Good ideas are often very old ideas.

Somewhere in the Upper Lower Devonian or Lower Middle Devonian, some group of nautiloids (opinions differ rather strongly *which* group of nautiloids it was) gave rise to a modest little group of coiled cephalopods—the ammonites. They were to become very successful as the Devonian wore on and gave way to the Upper Paleozoic. But their greatest successes lay ahead, for they became extremely abundant and varied in the Mesozoic, which is, in our consideration of life's history, the best place to review their rather incredible evolutionary story. But we should remember that from the Devonian onwards, ammonites were prominent members of the marine carnivore guild. They differed from nautiloids in the nature of their sutures—the wavy lines made by the intersection of the internal partitions and the outer walls of the shells. Nautiloids typically have very simple suture patterns (with rare exceptions), while ammonites usually have much more complex sutures (again, with rare exceptions). And ammonites almost always had a siphuncle running around close to the outer, bottom edge of the chamber—while in nautiloids of all kinds, the siphuncle always ran smack through the center of the shell, just as it does today in the pearly nautilus. The one exception to *that* generalization is that once again there was an early group of ammonites which did things exactly in reverse of what later became the standard theme for ammonites: these mavericks had their siphuncles running around close to the *top* of the internal chambers. It's either an example of an idea that didn't work—or more simply, just another group that fell victim to extinction before they had a chance to diversify, another way of saying "bad luck."

Ecological Systems of the Paleozoic

Thus the complexion of ecological systems in the Paleozoic was at once familiar and strange. It was familiar because the same basic economic activities we see going on around us today were very much in evidence back in the Paleozoic period. But it is also somewhat strange because the cast of characters carrying on the game of life back then were on the whole rather different from the creatures alive today. Economically speaking, there were "primary producers," the algae that photosynthesized and formed the base of the food chain, just as they do in today's oceans. And there were grazers, those animals that filtered the bacteria and ate the algal seaweed. There were floaters—plankton that included, in addition to the microscopic organisms, the large filter-feeding colonial graptolites. Graptolites, thought to be our own remote relatives, are particularly common in Ordovician black shale deposits, and have thus figured prominently in the correlation of these otherwise sparsely fossiliferous rocks. "Graptolite" means "written stone"—most graptolite fossils are carbon smears on shale surfaces, looking very much as if they had been pencilled in by some unknown hand.

And there were bottom herbivores—particularly all the snails that were crawling around back then (we will spend more time with the Class Gastropoda, Phylum Mollusca—in other words, snails—when we discuss the Mesozoic and Cenozoic, during which they became even more prominent in marine communities). There was the vast guild of sedentary food filterers, the creatures that have figured so prominently in this chapter. There were generalized scavengers, such as trilobites. There were the carnivores—animals that eat other animals—that we have just met. Thus the tempo, pace, structure, and fabric of life, of *economic* life, was very much as we see it today. The old theme of procuring energy and using it to grow, maintain simple bodily existence, and reproduce has been one of the two

FIGURE 42 A dendroid graptolite. These colonial organisms, apparently our own distant kin, were floaters in the surface waters of the inland seas of the Middle Paleozoic.

essential aspects of life since the beginning—even though the cast of characters filling all these roles was by no means the same as we have with us today.

And this sense of sameness within the Paleozoic is its most arresting feature: the survival of species for millions of years and the survival of entire *ecosystems* for millions of years. On every scale—from the appearance of new species up through the development of entire faunas and floras—we find an episodic pattern predominating: new species, new ecosystems appear relatively suddenly, more often than not replacing forerunners that had been disrupted or eliminated by extinction.

We shall follow the extinction-proliferation cycles through life's history, and consider the overall phenomenon in greater detail in Chapter 7. Now we must consider a somewhat different evolutionary effect: what happens to creatures when they are confronted with utterly new ecological terrain? For that is the central theme of life's conquest of the land, a story to which we now turn.

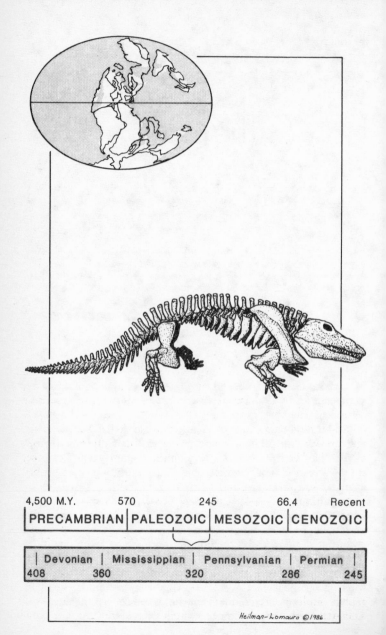

4,500 M.Y.   570   245   66.4   Recent
PRECAMBRIAN | **PALEOZOIC** | **MESOZOIC** | **CENOZOIC**

| Devonian | Mississippian | Pennsylvanian | Permian |
408   360   320   286   245

Heilman-Lomauro ©1986

5

Invasion of
the Land

For all Egypt's storied archeological ruins and paleontological treasures, most of the country remains barren desert. Yet even in the vast wastes of the western desert there is the occasional fertile oasis, and the odd remnant of bygone times. At Aswan, over 400 miles south of Cairo, now the site of the old Aswan Dam and the much newer Aswan High Dam, the ancients encountered what they called the "First Cataract." Precambrian granite abruptly rises through the overlying sediments and stands exposed in the river bed as impassable knobs of stone and gentle islands.

Exposed granite on shore provided a much-used quarrying site for Egyptian artisans and builders for well over a thousand years. Yet, passing by these arresting scenes, southeast of town, we still see the granite sticking up out of the ground. But now we are in desert, far from the fast-flowing Nile and well out of the narrow strip of arable land to which the fertile valley has dwindled as we travel upstream from Cairo to Aswan. The desolation of that granite and sand topography is chilling. Seeing it for the first time, I thought of the moon, so barren and hostile did it seem. But on second thought, it is about as good a place as any on earth to start considering the history of land surfaces and the life it has supported—*our* sort of life. Back in the Lower Paleozoic, *all* of the Earth's landscape was as barren and hostile as that region outside of Aswan. Yet in the mid-Paleozoic, life managed to invade the land and actually help convert it to a livable abode in relatively short order.

Detecting Traces of Ancient Lands and Life Forms

There was, we assume, always some semblance of land. When, decades ago, the prevailing vision of the early Earth saw it as a fiery ball slowly cooling and finally differentiating into a dense crust that formed giant catchment basins (the oceans) and lighter areas that bobbed up to form continents, there was a related notion that continents grew by a process of accretion. New mountain belts form on the margins of existing areas, appearing first as island arcs; after millions of years, these island chains are compressed, thrust up, and incorporated into the welter of rock types that form the continent. And there is some indication that mountain belts do increase in age as we approach the continental interior: the Appalachians, though far older than the Rockies, nonetheless are juveniles when compared with the so-called Grenvilles, which form the Adirondacks and a linear belt running up from the Thousand Islands region of the St. Lawrence River, straight up the west coast of that river through much of the Province of Quebec and on up into the mainland portion of the Province of Newfoundland.

Most theorists believe that the Earth has remained relatively constant in diameter, and that there has always been some measure of relatively light granitic crust. That is why we think that there has always been land exposed somewhere on the Earth's surface.

Yet the actual evidence of land is hard to come by. Sometimes, particularly in relatively young rocks, we actually find *paleosols*—literally, "ancient soils." Sometimes the evidence is less direct—scratches revealing the scouring of bedrock by an ancient glacier, or an erosion surface testifying to subaerial exposure. Most often, though, the evidence is even more inferential: the apparent deposition of sediments in fresh waters implies streams and lakes, which imply a terrestrial setting; or the appearance of a distinct layer of coarse gravel implies the sudden rise of the land surface nearby, raising the rate of flow of streams and enabling them to carry coarser pebbles farther than ever. Visiting the famous Devonian fish locality on the Gaspé, I first mistook a striking bed of gravel, a truly coarse conglomerate, for a wall of pebbles and concrete I thought had been built as a sea wall against the erosive powers of the waves of the Baie des Chaleurs. Only after I walked far enough to see that the supposedly man-made wall become overlain with normal sands and silts (with the fossil fish in them—some 470 million years old!) did I realize I was dealing with a natural deposit.

So we think there was always land, but the direct evidence of its presence is seldom available. Gravity, wind, and rainfall (the latter two are themselves dependent on gravity) make it inevitable that land will be erased, eroded down to sea level. Only constant rejuvenation will keep lands poised above the waves. With the ubiquity of erosion there is small wonder that the land itself leaves little trace in the rock record. The sedimentary rocks that provide most of our decipherable Earth history are areas where eroded sediments collected; the exposed land surfaces they came from were destroyed.

And that is why it is so hard to find the fossils of terrestrial organisms. Animals and plants living in streams and especially lakes are one thing: they are likely to die there, and stand a chance of being buried before their flesh decays and their bones become completely scattered. Creatures living near bodies of water, perhaps visiting them daily to feed, or simply to drink, are also likely candidates for fossilization. The early fossil record of the human lineage is especially well developed in the lake sediments of modern-day East Africa. Along Lake Turkana, for example, in the Rift Valley of northern Kenya, lie deposits known as the Koobi Fora Formation—filled with land-dwelling mammals that happened to die at the lake. The fossil collection includes occasional specimens of our ancient kin *Australopithecus*; our forebears were as dependent on ready access to water as we are.

But many terrestrial organisms live far removed from permanent lakes and streams. Forest dwellers are soon consumed by the hordes of microbes, fungi, and insects inhabiting the litter of the forest floor. The dry and dusty plains, deserts, and tundras are similarly unlikely to leave a rich fossil record of their biological inhabitants. The famous Flaming Cliffs of Outer Mongolia, where an expedition from the American Museum of Natural History discovered not only complete skeletons of the small dinosaur *Protoceratops*, but also some of their undisturbed nests with unhatched eggs still neatly arrayed inside, is a mind-boggling aberration. Covered by windblown sands and miraculously only exposed by the erosive winds and rains of the Gobi during our own era, these remarkable fossils remind us of the riches of past life that have gone without a trace.

So much for the bad news: life, born and raised in the sea, and thus finely attuned to an aqueous existence, had a hard time getting

FIGURE 43 Mounted specimens of *Protoceratops*, shown guarding a nest of eggs excavated at the Flaming Cliffs of Shabarakh Usu in Outer Mongolia.

started in the hostile environment that was land back in the Paleozoic. And land-living life stands even less of a chance of being preserved in the fossil record than do the brachiopods, clams, trilobites, and porpoises that have been populating the seas all these eons. Yet there is much to see in the fossil record of terrestrial life, a story that began sometime in the Upper Silurian, around 400 million years ago. And we are irresistibly drawn to that story, for it is, after all, our own story, too: it is simply a lot easier for human beings to grasp the meaning of life on land than to fathom what it is like to live attached to a coral reef fifty feet below the waves.

Establishing the Beachhead: Setting Up an Ecosystem on Land

I have spent many years off and on scrounging around quarries in Middle Devonian rocks in central New York State. The fossil hunting is usually pretty good—marine invertebrates are often exceptionally well preserved and common. But as a direct hint that something more than status quo marine community life was in progress, there is the odd piece of fossilized wood that hits the eye.

True vascular plants (meaning those that have tubes in their tissues to conduct water and nutrients up and down) are almost strictly creatures of the land. They are perhaps the most novel of life's terrestrial inventions. Eel grass, *Zostera*, alone among all true plants has managed to invade the seas—though there are many species of plants living happily in and around lakes and streams. But the surest sign that life was up and running on land is those bits and pieces of early wood that begin to show up in marine settings in the Devonian. Floating in from a nearby freshly upraised delta that was spreading westward across New York State, the wood chunks are an abrupt departure from more ancient days: paleontologists collecting in a similar setting in older Ordovician rocks, in a marine environment just offshore from another mountain system, have yet to find a single woody scrap. Sometime between the Upper Ordovician and the Middle Devonian, land plants put in an appearance and grew to sufficient numbers and size to start showing up as driftwood in shallow inland seas, well away from their newly found native habitat.

Plants are the basis, the *sine qua non*, of all terrestrial ecosystems. Life, *all* life, depends on one single external source for the energy needed for growth, body maintenance, and reproduction—just plain *living*. That source, of course, is the sun. It follows that the base of the energy supply in any ecological setting, whether on land or sea, resides with those creatures capable of trapping some of the sun's energy in the form of light waves, which they use to make their own food. Photosynthesis is a complex biochemical process taking place in specialized organelles within the cells of some single-celled microscopic organisms and in all but a very few of the multicellular plants. In the process, mediated by the presence of chlorophyll, sugar is synthesized from carbon and water, with energy from the sun's rays becoming trapped in the bonds linking the carbon and oxygen and hydrogen atoms together in the sugar molecule. All animals eat either other animals that eat plants or eat plants directly—one or more steps removed from the primary energy conversion provided by plants. Only fungi explore an independent path, deriving their nutrition directly through absorption of decaying plant and animal tissues. Obviously fungi, too, are ultimately dependent on the sun's energy; they merely have a novel parasitic way of getting at it.

To understand the functioning of a system often provides some immediate insight into the conditions that must have prevailed for that system to get going in the first place—as I have argued earlier in this book for our understanding of how life originally got started on the Earth. But sometimes the organization of a system is not such a reliable guide to its origins. And so it is with the invasion of the land: viewing the great economic importance of plants, it seems pretty obvious that they must have gotten out there first. Plants presumably had to establish an economic beachhead before the insects, vertebrates, and relatively few other animal groups that have made it to land could follow along.

Well, yes and no. The famous anatomist and paleontologist Alfred S. Romer once wrote a scenario of early life on land for the vertebrates—a story that made a lot of sense and was enhanced by a neat, paradoxical twist. Romer (who loved to tell stories anyway) speculated that the first amphibians to crawl out on land were barely much more than lobe-finned crossopterygian fish—relatives of coelacanths and lungfish, which were discussed in Chapter 4. Modern amphibians, including frogs, toads, and salamanders, retain the primitive "fishy" mode of reproduction: eggs must be laid in water or, at the very least, in moist places to ensure development. The eggs lack the refinement of an *amnion*—one of a complex of tissues ("fetal membranes") inside the egg that facilitates nourishment and provides protection to the developing embryo from the surrounding yolk mass. Amphibians also lack the body protection afforded by the scales typical of "reptiles" such as snakes and lizards, and present, too, in modified form in birds (feathers) and mammals (hair). Amphibians turn out to be another hodgepodge group: they are all land-living vertebrates that lack amniote eggs and scales.

Given the ties that frogs and other amphibia still have to the water for their reproduction and to keep their skins moist, theorists have long speculated that the earliest forays of vertebrates on land never took the beasts very far afield. The scenarios generally depict such creatures as the Upper Devonian *Ichthyostega* (known best from deposits in Greenland) living a life much like crocodiles today: basking in the sun along the banks of streams and lakes, but returning to the water for reproduction and perhaps to feed as well.

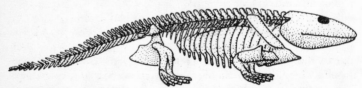

FIGURE 44 *Ichthyostega*, among the first of the land-living tetrapod vertebrate animals.

And here we find Romer's interesting, more detailed speculations coming to the fore. Romer knew that all these early land-living creatures had the jaws and teeth of carnivores. Nothing resembling a herbivorous vertebrate shows up in the fossil record until the later Paleozoic. But then, what was the advantage to going out on land at all—if reproductively, physiologically, *and* economically amphibians remained so utterly tied to the water? It didn't seem to make any sense.

Romer used to disparage his abilities as a field paleontologist. After all, he was a laboratory anatomist (his book on vertebrate anatomy has been pored over by countless premedical students over the years). Romer never thought of himself as a geologist, though he spent many seasons in the field, particularly in the Upper Paleozoic Red Beds of the southwestern United States, and also in Mesozoic rocks around the world, successfully bagging a plethora of fossils. He also made original geological observations on the rocks housing those specimens. The geological literature plus Romer's field experience led to his own particular version of "why the vertebrates decided to go ashore."

Romer saw that most of the sediments yielding the oldest amphibians, the earliest land-living vertebrates, creatures such as *Ichthyostega*, were *red*. In England and Scotland, it was the "Old Red Sandstone." There were red beds in Norway, Greenland, and the United States: the sandstone of the Catskills is red. All are about the same age, too: Upper Devonian. In the days before plate tectonics (or continental drift) was unreservedly accepted by the majority of practicing geologists, there was a notion of a sort of super "Old Red" continent that explained the similar appearance of these rocks over so much of northwestern Europe and northeastern North America. Nowadays it has become clear that there simply was no North Atlantic Ocean in Devonian times, that in fact the two continents were pushing closer to one another, and an enormous mass of mountains was being cast up along the suture between them—instantly shedding vast quantities of muds, silts, and sands that went coursing down to the shallow seas that flooded much of the continental interiors in those days.

The red color comes from the highly oxidized state of the iron in the sediments. (We have already seen in Chapter 2 how red beds are a tipoff of the presence of oxygen 2 billion years ago.) To a lesser extent, the red also comes from pink (orthoclase) feldspars, and the clays to which they commonly weather. The feldspars are common constituents of granites, which were being emplaced in the roots of this old mountain system. But Romer, following the conventional geological wisdom of his times, understood that the red implied actual desert conditions on the old supercontinent. Romer assumed that there was a progressive drying out of that early land environment; he visualized crossopterygian and other sorts of fishes caught in a nightmare of shrinking water supply. Lakes became ponds, ponds dwindled to pools. Competition was severe. In such stressful times, Romer proposed, it made a great deal of sense to get out: in his opinion, the first amphibians clambered out onto land, ironically enough, in search of bigger lakes, ponds, and streams—just as crocodiles and hippos vacate mud holes in search of permanent water when their seasonal homes dry up each year. It was not to get out on land because land promised something enticingly new: land was very much a hostile place. But it is conceivable that land could be seen simply as a *barricade* between a fish and a more inviting watery home.

It makes a good story—though, as I have already hinted, the question of whether the climate was actually drying up in Upper Devonian times remains debatable. Whatever the precise truth may be, crossopterygian fish seem to have been pretty well suited, in advance, to make that trip, to take the first few halting steps onto land. They had lungs as well as gills—a fortunate happenstance directly conducive to dealing with the major shift in oxygen extraction, from taking it out of water to getting it from the air. Fish "drown" in air every bit as readily as tetrapods (amphibians, reptiles, birds, and mammals) drown in water. Advanced bony fish have converted their lungs to air bladders to aid in buoyancy—thereby effectively "deciding" never to make a serious attempt to colonize land (speaking metaphorically, or course). The earliest vertebrates had gills and lungs, it seems, because they were gulping air, or perhaps simply holing up in mud wallows in dry times, as modern lungfish still do. And their lobe fins made rather ideal direct precursors for tetrapod limbs.

All of which is to say that the earliest amphibians were really crossopterygian fish with legs—just as the earliest automobiles were little more than motorized buckboards. With cars, there simply was no thought that a motorized four-wheel land vehicle could be construed as a wholly different sort of affair from a horse-drawn wagon. Entirely novel construction styles, from basic materials through fundamental aspects of design, are possible, but usually come after the first "experiments" have been performed. The first car is a motorized buckboard, the first tetrapod is a walking "fish."

Romer's scenario, even if it turns out not to be completely accurate, paints a rather different picture. Nothing like progress enters into the picture at all—the fish, remember, were just trying to get to water. They were "preadapted" in some important respects to survive brief exposure to air, thus to survive brief forays on land. And they had fins that could be turned into sturdy limbs, not all that dissimilar to substituting an internal combustion engine for a horse. Once out there, some vertebrates would inevitably begin exploiting whatever resources were available to them in that hostile territory. Perhaps at first they merely preyed on other species also wandering around in search of water. One thing is certain: vertebrates made the economic switch before they were able to break their reproductive ties to water. By the end of Devonian times they were integrated into communities; some had become vegetarians, and the rudiments of terrestrial food pyramids were already in evidence by the end of that period. But as far as the evidence suggests, the amniote egg—marking reproductive independence from water—came later, as much as 50 million years later. That's when reptiles first appeared.

Insects, Spiders, and Pill Bugs: Separate Arthropod Invasions of the Land

How do we measure evolutionary success? Perhaps we could use some measure of evolutionary longevity. If a particular species survives for, say, 15 or 20 million years, more than the 5 to 10 million years that is nearer the average duration of marine invertebrate species, there might be some justification in proclaiming its "success." Or perhaps we should look at the longevity of an entire group—in which case bacteria win, and sponges and coelenterates are bigger success stories than mammals (though the final results are still not in as long as they all continue to exist). Yet another way to look at it is to count up sheer numbers—number of organisms within a species, or number of species within a particular group of related species. If we choose that way to look at the matter, insects would win hands down over any other group of multicellular animals. From a sheer numbers point of view, insects are the very crown of creation.

It wasn't always like that: insects had their humble origins along with everything else. They first show up in the fossil record late in the Lower Devonian, in some rather poorly dated Scottish rocks known as the Rhynie Chert. And no one knows for sure just what group of aquatic Paleozoic arthropods gave rise to the insects. What is known is that millipedes, centipedes, and insects share a number of peculiarities that combine to convince entomologists of their shared evolutionary origins; wherever they came from, they appear very definitely to have descended from the same common ancestor. Beyond that, the range of theories on insect affinities is almost bewildering.

If we step back a moment, we can see that arthropods—that huge phylum of animals that includes trilobites, crustaceans, insects, and spiders and their kin—successfully invaded the land at least three times. How do we know this? Because we see the centipedes, millipedes, and insects forming one group. And we see spiders, scorpions, mites, and ticks—with their four pairs of legs and their unique pincers ("chelicerae") up front—forming a second, rather obvious and coherent terrestrial branch of arthropods. And we see a third type of land invader: the garden pill bugs (or sow bugs, as they are also called)—those little armadillo-like "bugs" that roll up when touched, and look for all the world like trilobites when unrolled and walking about in the grass of your backyard. They're not bugs at all (true bugs are an order of insects), but really crustaceans. They belong to the crustacean order Isopoda, very common in today's oceans.

And it is possible that other arthropods also made it onto land—only to become extinct without leaving a trace of themselves in the fossil record. In any case, all arthropods confronting the exigencies of life on land face a common set of problems, and all are equipped with the same general anatomical structures that in some instances abet and in others hinder that move to life in fresh air.

Take the simple matter of breathing. Crustaceans and the sole surviving aquatic member of the spider clan—horseshoe crabs—extract oxygen from water with filamentous gills. (It seems pretty sure that trilobites did the same, but the point is argued.) The gills soon dry up on exposure to air, though horseshoe crabs can survive for days with just a little moisture clinging to their protected gill surfaces. Gills are simply useless for removing oxygen from air—true for arthropods as much as it is for vertebrates. All land-dwelling arthropods—insects and kin, spiders and kin, and sow bugs—have hit upon the same basic solution, one for that matter hit upon by all land-living animals: they have, independently, invented "lungs." The internal anatomy of these lungs is recognizably different in each of the three groups; in each, air comes into the body through holes called "spiracles." Because these structures are not that similar, anatomists have long recognized that the "lungs" of all three had separate origins. Insects in general seem more similar to crustaceans, and arachnids more similar to horseshoe crabs, while terrestrial isopods have obvious relatives still living at sea. Thus few zoologists have ever thought that terrestrial arthropods formed one nice, coherent evolutionary group.

The independent evolution of "lungs" at least three separate times in arthropods once again illustrates the evolutionary phenomenon of "parallelism," in which similar adaptive solutions are fashioned to handle the same basic engineering problem—in this instance, extracting oxygen from the air. Holes had to be punched through those tough outer skeletons—otherwise so well suited to seal off the internal workings from the desiccating action of fresh air (a problem far easier for the arthropods to deal with than for the vertebrates). Because evolutionary theory ever since Darwin has in fact been almost solely a theory of the origin of adaptations through natural selection, it was not particularly difficult to imagine arthropods conquering the land on three separate occasions—a fact abundantly confirmed by the apparent separate evolutionary affinities of insects, arachnids, and isopods to other groups of arthropods, some still living beneath the waves.

But the "principle of parallelism" has been of late carried to a ridiculous extent in arthropod studies—to the point where it has become fashionable in some circles to call the arthropods a hodgepodge "grade," a simple collection of adaptively similar organisms whose joint possession of what appear to be the same basic features merely reflects the independent attainment of the same basic engineering "solutions" to similar problems of living. Thus the centipede-millipede-insect group, it is claimed, was independently derived from some worm-like group—independently, that is, from crustaceans, trilobites, and arachnids. It is not uncommon to read theories of three or even four separate origins not of *terrestrial* arthropods, but of arthropods in general. Indeed, to verge on the extreme, astronomer Fred Hoyle and mathematician N. C. Wickramasinghe, in support of their theory of the origin of life on Earth from cosmic seeding from outer space, have picked up on recent arguments on the separate origins of arthropods, claiming that each group sprang from a separate seed, and that, moreover, the dissimilarities even between different orders of *insects* indicate that they all came from separate seedings. The order of the various groups' appearance in the fossil record simply means, to these savants, that life has come from outer space in distinct waves of invasion.

Well, most claims on the separate origins of arthropod groups have not been as downright silly as the Hoyle/Wickramasinghe fantasy. But they do show what the allure, the attractive power, of natural selection as an explanation for all manner of phenomena can be for the unwary. There are several features, particularly of the head and specifically in the jaws, that indicate close affinities between insects and crustaceans. Indeed, there was a group, the "Mandibulata," long recognized as containing the two together as a distinct, unique evolutionary branch of the arthropods, and zoologists seem to be returning to that interpretation of insect origins that they are affiliated with the older, marine Crustacea. And in general, evidence from the biochemistry of the exoskeleton alone suffices to show that arthropods as an entire phylum are truly a single, natural evolutionary group and, by many a yardstick, the most successful phylum yet to hit Earth, whether by evolutionary or cosmic means!

Early Horseshoe Crabs: Tracking Their Life on Land

Some of the best fossil hunting is done these days in museums—rummaging around dusty, long-unopened drawers that house some forgotten or unappreciated gems collected decades ago.

Another excellent way to gather novel, exciting material is simply to sit back and let it come to you—which it will occasionally do if you happen to be in a place that people would choose to bring their specimens to be checked out—a place such as the American Museum of Natural History, for example.

One day in 1972 I received a call from a colleague, Bobb Schaeffer, then the American Museum's curator of fossil fish. An American missionary, LeGrand Smith, who was lving in La Paz, Bolivia, had stopped by to chat and show off a few specimens—including a baseball-sized lump of hardened mud that didn't seem to be much of anything. But Smith mentioned trilobites to Schaeffer, which was enough for him to call me down; and since it's always important to make contacts, I went.

I took a look at that "inorganic concretion" and saw what I took to be the lenses of a trilobite eye peering out of the matrix. Then I realized that arthropod-like skeletal material was also partially exposed, and my excitement grew. I spent the rest of the day trying to convince Smith to leave the specimen with me so that it could be properly cleaned—and studied and, if it was truly new, described in the scientific literature. To sweeten the pot, I offered to name the creature after Smith if, as I suspected, it couldn't readily be assigned to any known fossil already named.

FIGURE 45 The synziphosuran (primitive horseshoe crab) *Legrandella lombardii* from the Lower Devonian of Bolivia.

Smith did leave the fossil behind, and our technician labored diligently over it for nearly a month while I went out into the field after still more trilobites. (The technician, Frank Lombardi, did such a fine job that part of that fossil's name honors him: *Legrandella lombardii*.) When I got back, I was amazed: the head, cleaned off first, was not that of a trilobite, but rather of a primitive sort of horseshoe crab. Naturally we expected to find a solid shield behind the head, corresponding to the solid abdominal region of modern horseshoe crabs. But instead, Lombardi uncovered a series of articulated and very trilobite-like segments, culminating in a triangular-shaped spiky tail—just like the modern horseshoe crab tail (telson), and unlike anything encountered in a trilobite.

Legrandella lombardii enabled me to reinterpret the early stages of horseshoe crab evolution, clarifying many of the anatomical features of these primitive so-called synziphosurans (essentially horseshoe crabs with unfused abdominal segments), which had been known from relatively few and almost invariably poorly preserved remains in Silurian and Lower Devonian rocks scattered throughout the world. It also got me in touch with LeGrand Smith, whose extraordinary specimen led me to a long-term study of southern hemisphere trilobites.

It may seem a bit odd to go into *Legrandella* and its horseshoe crab relatives here, in a chapter on the invasion of land. Horseshoe crabs, however, have had their own flirtation with the land. True, *Legrandella* and its closest relatives were definitely marine creatures, coming from beds with brachiopods, trilobites, and other typical marine fossils. And modern-day horseshoe crabs are also saltwater beasts, though I have already mentioned how *Limulus polyphemus*, our North American Atlantic coast horseshoe crab species, can withstand subaerial exposure, surviving on occasion for several days. They crawl up onto beaches to excavate shallow nests in which to deposit their eggs—evidence indeed that horseshoe crabs are, for brief intervals every year, accustomed to life on land. They come, the females dragging the smaller males along on their backs, for reproduction, *not* for feeding, in a very real sense the opposite of the early vertebrates, in which reproduction was still tied to the water long after terrestrial ecosystems were up and running.

But there is more to the horseshoe crab/land connection than occasional, if regular, forays to the beach to reproduce. Daniel Fisher, now at the University of Michigan and perhaps the foremost student of horseshoe crabs these days, has done some exceedingly clever, innovative research on fossil and recent horseshoe crabs. In his initial work as a student at Harvard, Fisher etched specimens of the Upper Jurassic *Mesolimulus walchi* out of the Solnhofen

FIGURE 46 The modern horseshoe crab *Limulus polyphemus*. This one was comparatively young when it molted for further growth.

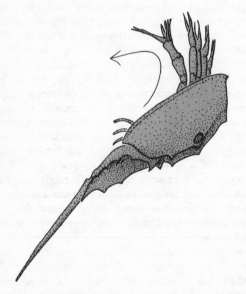

FIGURE 47 Horseshoe crabs swim on their backs, periodically shedding a vortex of turbulence (arrow) to provide a frictionless backstroke for the beating legs.

limestone of southern Germany. These specimens have long been taken as virtually identical to modern horseshoe crabs, though it is obvious they were somewhat flatter. Fisher made accurate wax models of the fossils, and sought to explain the slight but consistent differences in shape between the fossil and Recent species. He found that modern horseshoe crabs swim on their backs at an angle of about 20 degrees to 30 degrees. They achieve a speed of about ten to fifteen centimeters per second. By performing experiments in a tank, he soon hit upon the reason why they swim at that angle and speed: the animals use their legs as oars, rowing through the water. Since they cannot raise the "oars" above the surface on the backstroke (the way we do when rowing a boat), they need some other way to achieve a "frictionless" backstroke so that they will continue to move forward, and not just eddy back and forth as their legs beat steadily away. The horseshoe crab headshield is like a bowl; Fisher found that the deep cavity housing the legs periodically sheds a vortex of water as the animal moves ahead at a particular angle—between 20 and 30 degrees. The legs move forward with the vortex, which is then shed and the legs beat backwards once again, ready to snap forward with the next vortex. Fisher went on to show, with his wax models of the fossils, that the Jurassic fossils (assuming only that they too swam on their backs, utilizing the same vortex method with the backstroke) did so while inclined nearly horizontally, thus achieving a somewhat faster speed of 15 to 20 centimeters per second. All in all, a masterful piece of paleontological investigation!

Fisher then turned to the Upper Paleozoic record of horseshoe crabs. In the famous beds of Mazon Creek, in the soft coal region of central Illinois, collectors have been turning up enormous quantities of fern fossils, insects, and horseshoe crabs for well over a century. The fossils invariably come in concretions—oblong, generally flattened, yet rounded "stones" that, when cracked, often yield an exquisite fossil at their center. (Concretions form as mud wraps around an object, such as a pebble, twig, or invertebrate shell; *Legrandella* was similarly ensconced in such a mudball concretion.) In more recent years, a nearby and equally impressive saltwater fauna (the Essex fauna) has also turned up, containing, among many other prize rarities, occasional horseshoe crabs rather similar to our modern marine forms.

But the long-known Mazon Creek horseshoe crabs are something else again. First of all, they are in there with all those ferns and insects, indicating fresh- rather than saltwater. They were definitely in those coal swamps, not out in the saline lagoons or open seaways with the other type of horseshoe crabs. The generic name of these Upper Paleozoic coal swamp horseshoe crabs is *Euproops*, and they are known from similar environments in Carboniferous rocks in Europe as well as the United States. Fisher has shown that the solid trunk shield of *Euproops* and its closest relatives is constructed rather differently than the shield of modern *Limulus* (remember that solid shields were derived from the more primitive free-segment condition of the earlier members of the group), suggesting the distinct possibility that the *Euproops* sort of horseshoe crab evolved independently from the "main line" of horseshoe crabs.

In any case, Fisher believes he has both solid and circumstantial evidence suggesting that *Euproops* was not merely a freshwater beast, but actually a member in good standing of the terrestrial, coal swamp ecosystem. Specimens have actually been recovered from the fossilized stumps of an ancient upright lycopod tree: and Fisher also argues that the round shape of the *Euproops* abdomen, plus its long spines jutting from the rear margins of the eyes, were ideal camouflage as these animals crawled up and down spikes of vegetation on land. The latter, rather imaginative speculation, is unfortunately not buttressed by as airtight a series of experiments as Fisher's analysis of the way that marine horseshoe crabs swam, but it's nonetheless more than just a wild guess. I think that Fisher has actually shown that some horseshoe crabs formed a fourth arthropod group—beyond insects, isopods, and the spider contingent—that was successful, for a brief time at least, in making the transition from water to land. Adding up the inferences from modern-day marine horseshoe crabs, and Dan Fisher's analytic abilities and keen insights, we come up with a pretty convincing case that more sorts of arthropods walked the Earth in the past than do today.

Scorpions and Spiders

Scorpions and spiders, the better known terrestrial chelicerates (mites and ticks completing the roster), were already in existence in the Lower Devonian, and there is no hint that their invasion of the land had anything to do with horseshoe crabs also giving it a try, even though horseshoe crabs are themselves chelicerates—after the chelicerae, or pincers, that all of them bear as their forward-most pair of appendages. These chelicerates *are* close kin, so close that Ray Lankester, a famous British zoologist of the nineteenth century, wrote a terse little note in 1881 entitled, simply enough, "*Limulus*, an Arachnid." Modern classifications restrict the term "Arachnida" to spiders and close kin, seeing horseshoe crabs and other, extinct groups as related to spiders only within the Subphylum Chelicerata.

But Lankester's point was clear enough: *Limulus* shares a number of special similarities with spiders and scorpions, items such as those chelicerae, otherwise encountered nowhere else in arthropods.

All true crustaceans share some peculiarities of their own. For example, all have two pairs of antennae. No chelicerate, marine or terrestrial, fossil or Recent, has ever been observed with even a single pair of antennae. In recognizing natural groups on the basis of shared anatomical structures, Lankester was applying solid principles of classification that strike me as quite modern.

Scorpions pose one final problem as we contemplate arthropod invasion of the land. Though they are chelicerates, it is by no means a foregone conclusion that they have a particularly close evolutionary affinity with spiders. Indeed there is no trick to tracing the scorpion invasion of the land—because perfectly good scorpions were already well developed, alive and well and living in saline lagoons in the Upper Silurian of what is now New York State! Two fossils were discovered in the last century from beds of the Bertie Waterlime, otherwise justly renowned for their eurypterid fossils. Eurypterids, recall, are themselves very scorpion-like, and indeed are commonly called "sea-scorpions." The eurypterids, though true chelicerates, lack the stinger on the end of the abdomen—that *sine qua non* of all true scorpions. Nonetheless, it had always been rather easy to imagine that eurypterids had given rise to a land-living contingent of scorpions. But with the discovery of aquatic scorpions, the

FIGURE 48 An aquatic scorpion from the Silurian Bertie Waterlime of New York.

game changed a bit: scorpions needed simply to trade in their gills for lungs to make the trip out to land. Otherwise their external anatomy, at least, remained virtually unchanged from the days when scorpions still lived beneath the waves.

In fact, recently discovered deposits in Germany, along the Mosel, help paint a picture of a number of different lines of arthropods—mostly different chelicerate groups, but also the crustacean lines leading to insects, isopods, and perhaps other, now extinct groups—all out there making some use of the land. Some were marine, yet apparently clambered ashore at least some of the time, perhaps as modern horseshoe crabs still are wont to do. Others were up on shore, already making the transition from water to land. The literature registers strange names such as "diploaspids," enigmatic groups of chelicerates whose affinities with the better known fossil and living groups are still very much in doubt. We get a strong feeling of experimentation, an exuberance of life as a variety of lineages all took their shot at gaining a foothold in what, with the advent of photosynthetic life ashore, was literally a land of great opportunity. Nature truly abhors a vacuum, and where economic opportunities open up, and reproductive essentials can be maintained without disruption, life is sure to fill that vacuum. It is the well-known but always stunning phenomenon of *adaptive radiation* once again: when a new ecological arena appears, or extinction resets the ecological scene, so that replenishment is required to regain the full complement of life, new forms rapidly appear to fill the void.

First came the plants, but also, virtually simultaneously, the arthropods. The vertebrates were not far behind. All was up and running, as we have already seen, by the end of the Devonian, when one of life's more massive extinction events leveled many marine species, leaving a less noticeable effect on the younger ecosystems on land. Land snails appeared somewhat later. And this brief list of animals that successfully invaded the land is the complete roster. For though it is true that protozoans of many kinds, also various sorts of "worms," (as well as true, segmented worms) live in the actual earth (and are therefore very much "terrestrial" creatures), only a very few groups of animals were able to successfully colonize the land, to live in fresh air independent of water except for occasional visits to replenish body fluids and, for some, to reproduce. Life, after all, was invented in the sea, and the bodies of all organisms bear the unmistakable primitive stamp of being developed first to live in a watery chemical environment. Only a few of those body systems were able to be modified enough to withstand the rigors of life out of water.

Life Dichotomies in the Upper Paleozoic

From now on the ecological action is split: we must follow life on land as well as in the sea for the remaining 300 million years from the Devonian to modern times. And lineages also split, with some developing on land and some in the sea. Take, as a spectacular example, "reptiles."

On the basis of a skeleton alone, it is difficult always to be sure of what is a "reptile" and what is an "amphibian." As we have already seen, amphibians are really primitive tetrapods—land-living, four-legged (except those few that have become snake-like) vertebrates that lack scales and amniote eggs. Anatomists have settled on a few bony criteria to distinguish reptiles from amphibia in the Paleozoic, but in the absence of eggs (not at all preserved, so far as I know, in the Paleozoic) or information about scales, things are tough—the precise origin of amniotes in the Paleozoic is presumed to have coincided with creatures called "reptiles" based on the configuration of bones in their skulls. *Seymouria*, from the Lower Permian of Texas, was the old prototypical, original reptile, though later students have seen *Seymouria* as only a pseudo-reptilian, an "amphibian" that was advanced and well equipped for terrestrial life—but an "amphibian" just the same.

These Upper Paleozoic amphibians are really wonderful creatures. Some are almost fabled, particularly the labyrinthodonts, which include the wicked *Cacops*, *Edops* (known, colorfully, if somewhat cryptically, as old "Grandpa Bumps"), and my personal favorite, *Eryops*. These were squat, crocodilish sorts of beasts, always depicted with a rather baleful glare fixed on their reconstructed countenances. They must have been *the* prodigious carnivores of their times, which went on for quite a while. They didn't yield automatically to the advent of true "reptiles" in the Mesozoic, and labyrinthodonts are found through the Upper Paleozoic right up into the Triassic. Indeed, it was a labyrinthodont identified by Edwin H. Colbert of the American Museum of Natural History that clinched the presence of Triassic fossiliferous rocks in Antarctica. There were others as well in the roster of Paleozoic amphibians. Some were rather small creatures haunting those coal swamp habitats, while others were nearly as large as the labyrinthodonts: *Diplocaulus* and related genera were among the more bizarre animals on Earth, with their strangely delta-shaped heads and long skinny bodies.

But before we get to that great reptile split, a word of explanation is in order for those great coal swamps. Paleontologists in the United States refer to the Upper Paleozoic as a succession of geologic

FIGURE 49 Reconstruction of *Seymouria*, variously considered a primitive reptile or an advanced amphibian.

systems: Mississippian, Pennsylvanian, and Permian (after, as is usual, places where rocks of that age are especially well developed or first recognized; Perm is a small town in the western region of the Ural Mountains of the Soviet Union). In Europe, the Mississippian and Pennsylvanian coincide almost exactly with the Lower and Upper Carboniferous—named for the great coal deposits of Britain and the continental mainland formed from the great coal swamps so well developed at that time.

Yet coal is frequently encountered throughout the geologic column. The great strip-mining in the west (particularly in Montana and in New Mexico) is actually done in rocks of Paleocene age, the lowest subdivision of the Tertiary, lying above the dinosaur-bearing rocks of the Cretaceous. Coal is formed where plants growing rich and thick become buried in such quantities and in such chemical

FIGURE 50 *Eryops*, a Lower Permian amphibian.

circumstances that decay is slow and incomplete. The great masses of vegetation become compacted from the tremendous weight of the layers of sediment that continue to accumulate above them. The final mass is the dense, carbonized strata we mine and quarry for fuel.

Reptiles became abundant toward the end of the Paleozoic. Typical among them, and fairly common as fossils in the Permian-aged red beds of the American southwest, were the captorhinomorphs, dog-sized creatures with a skull roofed over with a continuous layer of bone. The latter point is important, for the formation of the skull, with or without holes, is what distinguishes all major kinds of amniote vertebrates—not just reptiles, but mammals and birds as well. Turtles today continue the captorhinomorph tradition—no holes in the side of the head. The other reptiles alive today—snakes and lizards, crocodiles, and New Zealand's tuatara—*all* have two

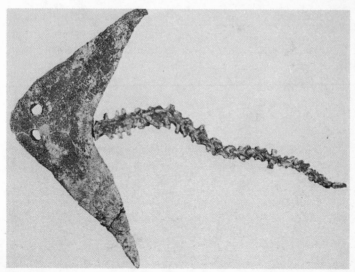

FIGURE 51 Head and vertebral column of *Diplocaulus*, an unusual Paleozoic amphibian.

holes on both the left and right sides of the head. They are "diapsids." Birds are also diapsids; in a very real sense, birds are simply advanced, feathered "reptiles."

What, then, of mammals? We, it turns out, are every bit as much "hairy reptiles" as birds are "feathered reptiles." Though nothing we would call a reptile remains on Earth today with our sort of skull configuration, by late Paleozoic times a major subdivision of reptiles with just our sort of skull had arisen, and indeed was dominating life on earth in several quarters of the globe, most notably and persistently in the Gondwana ecosystems that have left their fossilized remains in Brazil, Argentina, Antarctica, and the Karroo Basin of South Africa. These were the "synapsids," distinguished by their single perforation in each side of the skull. That perforation can be traced through a series of fossils up to the modern mammalian skull, with the hole now enlarged and covered over by the so-called "zygoma," which persists on our own skulls and connects our cheeks with the rest of the skull.

Thus we have several splits with which we must contend: life on land and sea; amniote vertebrates embarking on two major (there were others in the old days) evolutionary pathways, one that led, eventually, to birds, and the other toward mammals. For now we will follow the therapsids (a division of the synapsids, usually called "mammal-like reptiles") on land, and see how the Paleozoic ended at sea. We can then turn to the Mesozoic, where life on both land and sea underwent something of a major overhaul: mollusks and bony fish come to dominate the seas, and the archosaurian diapsids (dinosaurs and related beasts) come into their own on land.

Upper Paleozoic Life in Texas

There is at least one university in the world that bears the name of a geologic period: The University of Texas of the Permian Basin. I've

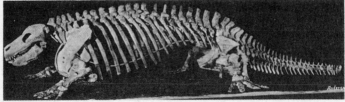

FIGURE 52 *Diadectes*, a Lower Permian reptile. Like turtles and captorhinomorphs, *Diadectes* belonged to a group that lacked openings on the sides of the skull. Such openings, variously positioned, are diagnostic of the lineages that led separately to birds and mammals.

always loved that name. Located in the "Permian Basin" of southwestern Texas, the campus reflects perhaps the most important attribute of the region: one of the finest developments of Permian rocks to be found anywhere in the world. Few other regions take their name from geology and paleontology, though the area just south of Prague (Praha) in Czechoslovakia is still known as the Barrandium, after the great nineteenth century paleontologist Joachim Barrande (1799-1883), who described so many of the exquisitely preserved Paleozoic invertebrates from the region in huge, quarto-sized monographs liberally illustrated with drawings of all the fossils.

The Guadeloupe and Glass Mountains of southwestern Texas expose great thicknesses of limestones and shales. The fossils discovered in the 1800s by geologically minded explorers could not be recovered and studied to full advantage until the great brachiopod paleontologist G. Arthur Cooper (now retired, but still working indefatigably at the National Museum of Natural History, Smithsonian Institution) developed a technique for massive etching to free the fossils from their tough surrounding matrix. Many of those Permian brachiopods, clams, snails, sponges, and corals, are *silicified*—meaning that the calcareous shell is replaced entirely by silica (SiO_2). Limestone breaks down to carbonic acid and water when exposed to hydrogen chloride (HCl, hydrochloric or muriatic acid):

$$CaCO_3 + 2HCl \longrightarrow H_2CO_3 + CaCl_2.$$

The reaction is irreversible, because the carbonic acid (H_2CO_3) readily breaks down to stable molecules of water and carbon dioxide, which is readily given off as a gas:

$$H_2CO_3 \longrightarrow CO_2 + H_2O.$$

Dipping a chunk of relatively pure limestone into even a mild concentration of muriatic acid produces a vigorous reaction, and care must be taken lest the carbon dioxide bubbles are produced at such a violent rate that the delicate silicous fossils are broken up in the roiling liquid.

Cooper set up an etching lab at the Smithsonian and began the mammoth task of removing all those fossils from tons of rocky matrix. The results in terms of brachiopods alone is mind-boggling: Cooper and his colleague Richard Grant have written six large monographs describing hundreds of brachiopod species from the Permian of Texas. The feat is prodigious in human terms, and highlights both the tremendous diversity of brachiopods *and* the benefits we get from extraordinary circumstances of preservation that allow us added access to the treasures of the past.

Cooper focused on the brachiopods, and a generation of graduate students concentrated on the other important groups of fossils in the same beds, fossils that came out in huge numbers along with the brachiopods. Virtually all the students who wrote doctoral dissertations on these fossils in the 1950s were students of Norman D. Newell of the American Museum of Natural History and Columbia University. Newell, himself an expert on Paleozoic clams, had a deeper scientific problem in mind beyond the task of supervising the analysis and description of all the different kinds of fossils. Newell was interested in paleoecology (the economic relations among animals and plants in bygone settings) and he had decided to focus on reefs and associated environments. He developed a two-pronged approach: with some colleagues and students, he initiated some very important research in the living communities of the Bahama Banks, studying the composition of the faunas in different environmental settings as well as the mode of formation of the calcium carbonate—*limy*—sands that litter the bottom of Bahamian waters. And with the support of the oil industry, he initiated a reconstruction of the ancient geography and environments of the Delaware Basin in Texas. He was looking, of course, to apply principles gleaned from a knowledge of the Recent scene to understanding the past, but he kept a weather eye out as well for lessons that the past might have for our grasp of the modern ecology of the marine environment. Newell, more than anyone else, has given us a comparative approach and an appreciation of the development of reefs through time.

The picture of what the now dry and dusty Delaware Basin looked like so long ago is magnificent. El Capitan, a conspicuous promontory of the Glass Mountains, was once a massive reef front. Behind it stretched a shallow lagoon, with its own characteristic suite of organisms and its tendency to become cut off from the open, normally saline waters—to the point where salt would begin accumulating from time to time. But in visualizing this ancient Permian scene, it is hard, as one drives up to that ancient reef face across the depressed region of the Delaware Basin, to avoid the supposition that erosion has etched away the surrounding rock, leaving behind the harder, more resistant reef rocks that now form El Capitan. And that supposition would be right—over the last 250 million years, sediments were deposited in that basin, and have been eroded away. But Newell and his colleagues and students realized that the rocks which floor the modern Delaware Basin *are the same age as the reef rocks which tower above the Basin*. In other words, the modern-day relief directly reflects the ancient scene: a huge reef faced outwards on a deep basin—a basin, moreover, that housed a

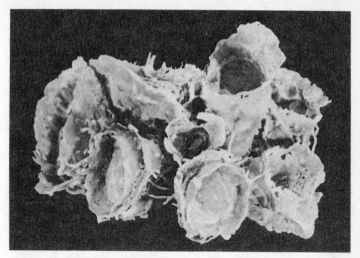

FIGURE 53 Specimens of *Prorichthofenia*, a bizarre articulate brachiopod. Inhabiting reef environments in the Permian (now the southwestern United States), this brachiopod mimicked corals with its deepened pedicle valve and the much reduced brachial valve utilized as a simple lid.

stagnant, oxygen-poor and rather stinking seafloor unfit, for the most part, for normal marine life. Ammonites and other floaters and swimmers are the most common fossils out there, creatures that lived and died above the seafloor.

More Evolutionary Scenarios: Permian Vertebrates of Texas

The varied environments of the marine Texas Permian is only half the story of the paleontological picture of life in that corner of space and time. Land was close by the Delaware Basin, supporting a rich array of terrestrial life. The Upper Paleozoic red beds of the Southwest have long been a productive source of vertebrate fossils. Best known, perhaps, were the rather vicious-looking synapsid pelycosaurs, among the earliest of the "mammal-like" reptiles.

One of the weakest links of evolutionary biology, as I've pointed out before, has been the tendency to attribute all manner of power to natural selection, the force that indeed does mold organisms to their environment. There really is design in nature—organisms on the whole tend to be well suited, or adapted, to their habitats. And relative ability to survive in the wild has an obvious corollary in the probability that an organism will be able to reproduce: in general, those organisms that thrive in their environment will tend to leave more offspring than organisms not as well endowed for the economic game of life. So there is no doubt that natural selection is a powerful force—and *the* force that shapes and maintains the adaptations of organisms.

But it is one thing to accept natural selection as the molder and shaper of organic adaptations, and another to understand the true nature of an organism's form and to understand how that form came into being. There has been a tendency in science to make up stories about "how the giraffe got its long neck"—and simply invoke natural selection as the all-powerful machine that can do the job regardless of how implausible a particular scenario might be. We have already seen how instructive some such stories may be, such as Romer's scenario, recounted earlier in this chapter, on "how the vertebrates came out on land." But some of the stories are simply ludicrous—none more so than a tale arising out of pelycosaur anatomy and Permian paleogeography in the ancient Southwest.

Dimetrodon was a vicious-looking pelycosaur that attained a length of some eleven feet. Moderately common in the Permian red beds of the Southwest, *Dimetrodon* came with a rather narrow and deep skull replete with a full complement of wicked-looking, sharply pointed teeth. Unlike nearly all of its relatives, *Dimetrodon* developed vertebrae along the back into extremely long spines. In life, these spines would have been covered by a layer of skin and scales, amounting to a large, rounded, sail-like affair that would be an oddity for any vertebrate living or dead. It is this "sail" that has prompted so much speculation, including my favorite "Just-So" story in all the annals of paleontology.

It seems another pelycosaur was more common in the Permian of the region: *Edaphosaurus*. Like *Dimetrodon*, *Edaphosaurus* grew to about eleven feet in overall length, and was also equipped with long spines along its back. It, too, had a "sail." The only difference between the sails of these two genera—and it is a big difference, as it enables instant identification of scraps of bone in the field—is that there were little cross-twigs of bone jutting out from the long spines in *Edaphosaurus*, while the spines in *Dimetrodon* were entirely smooth. Question: what on Earth was the function of those spine-supported "sails" on their backs?

Answer: no one knows. Or, as Romer put it in his textbook, "As good a guess as any is the suggestion that it was a heat-radiating device which may have aided in the control of internal temperature"—another moderate, plausible speculation. But best of all, for its sheer lunacy, was the idea that since *Dimetrodon* tends to be found in sediments on one side of a marine basin, and otherwise rather similar pelycosaurs that *lack* such a sail occur in rocks on the other side, that *Dimetrodon* and other "sailed" pelycosaurs were actually males that literally used their backs to sail across the bay to their lady loves! Here is a principle—sexual dimorphism (meaning that males and females are often consistently different in one or more respects)—taken to the *n*th degree and mixed in, for good measure, with the accidents of collecting and the known facts of ancient geography to yield a silly scenario. Although it is sometimes true that organisms go to desperate lengths to procreate, there is nonetheless

FIGURE 54 The Permian *Dimetrodon*, a vicious-looking sail-backed synapsid reptile.

little doubt that once again Romer's is the more reasonable view: assuming there to have been a function for those sails, it is most likely that it was *economic*—perhaps, as Romer suggests, related to the problem of controlling body temperature. If it were sexual, it is far more likely the sails were species recognition marks—like the various twisted spirals of African antelope horns—rather than an actual physical mechanism that allowed the males to reach the females. Paleontologists really must be guided by the biological systems still alive as their main mode of inference for behavioral lifestyles of the past!

The End of the Paleozoic, Part I: The Great Dying

Life was shaken up rather abruptly at what we now recognize as the Permo-Triassic boundary—the same line of demarcation that serves as the great divide between Paleozoic ("Ancient Life") and Mesozoic ("Middle Life"). I have already remarked that extinction events are real, and perhaps the very best evidence of extinction was the tendency of pre-Darwinian, creation-minded geologists to divide up geologic time according to natural divisions of life as seen arranged vertically in the rocks: phenomena we now know to be relatively sudden extinctions followed by equally large-scale proliferations. Some extinctions were greater than others, involving more different kinds of organisms over large regions of the Earth's surface. The extinction that wreaked such ecological havoc at the end of the Paleozoic was probably the most devastating to have hit life so far. Paleontologist David Raup of the University of Chicago estimates that over 90 percent (perhaps as much as 96 percent) of *all species* became extinct in the relatively brief span of roughly a million years. Yet even at the close of the Permian, the extinction was by no means uniform: some groups were harder hit than others, while some regions were less hard hit than others.

What caused the demise? Seas had been withdrawing from the land surfaces as glaciers grew, reducing habitat area and indicating the onset of cold conditions. Two general choices are available to explain these catastrophic collapses of regional, and even worldwide, ecosystems: either such environmental change builds to a critical point where increasingly bad conditions become intolerable and the whole system is imperiled; or, as has become increasingly popular in recent years, disturbances outside the system, for example, an asteroid colliding with the Earth, might well be the culprit triggering the devastation. Most hotly debated these days has been the only slightly less catastrophic extinction that ended the

Mesozoic, when dinosaurs, ammonites, and a host of other creatures bit the dust (perhaps literally if it really was an asteroid that got them). Because mass extinction is a recurring phenomenon throughout the history of life—and a dominant source of rhythm in evolution—I will develop a general discussion of extinction in a chapter of its own (Chapter 7), following our look at Mesozoic life. Like most paleontologists, I have my own opinions on the topic of mass extinction—a sure sign of activity in scientific ranks.

The End of the Paleozoic, Part II: The Scene in Gondwana

But there were *two* themes to the end of the Paleozoic. Sudden, massive extinction is one. Relative continuity is the other. We know now that Gondwana was a united mass of South America, Africa, India, Australia, and Antarctica. Part of the continuity of life in the Upper Paleozoic story derives from the close contact of these continents for so long. In the Southern Hemisphere, conditions had amerliorated somewhat from the days of the Devonian, when the South Pole was perched not far north of present-day Cape Town. With a more temperate climate and plenty of lakes and streams, terrestrial life flourished, leaving in its wake a rich and varied fossil record. *Mesosaurus*, a rather enigmatic little reptile with a set of wicked-looking teeth presumably used to spear fish, has long been a point of contention in the age-old dispute over "continental drift"—that is, until the resolution of the problem in the 1960s when the Earth sciences as a whole experienced an intellectual revolution and openly embraced the notion of shifting continental land masses. Since well back in the nineteenth century, some scientists have realized that the distributions of animals and plants indicated rather close connections between some continents that now are rather far apart. The Southern Hemisphere continents are the most obvious. Many similarities in plant and animal life indicate a common origin separate from related forms to the north.

Yet geologists for the most part were "stabilists," and paleontologists had to be content with long-range migration, often over imaginary land bridges connecting far-flung continents, to explain these distributions. Indeed, it was the animals and plants that provided the so-called "evidence" for these "land bridges." *Mesosaurus* is known especially well from a distinctive Carboniferous horizon in South Africa. But virtually *identical* specimens have also been recovered from rocks of the same age in Argentina. Now, the rocky matrix that produces *Mesosaurus* skeletons offers fairly clear evidence that they led a freshwater life. There is no way *Mesosaurus* could have traversed the South Atlantic by swimming. So there *must* have been a land connection, complete with lakes, between Africa and South America in Carboniferous times, or so argued the stabilists. But *Mesosaurus* was such an egregious example that all the "crackpot" diehards who still insisted on continental drift always cited that distinctive beast as evidence of drift, and *Mesosaurus* became such a symbol of a "lost" cause that it was a standard exam question in the days just before drift became respectable.

The theme of a partially isolated Gondwana ran right across the Permo-Carboniferous boundary. Extinctions down there were not nearly so severe as they seem to have been in other reaches of the globe, nor were they even exactly correlated with events in the north, at least in the opinions of some paleontologists who have looked at the question. By far the best sequence is smack in the middle of old Gondwana, in a huge basin exposed in the Karroo desert of South Africa. The rocks, with a sequence of some five major faunal zones, are called the "Karroo" system—a body of rocks and a coherent, if changing, fauna that bridges the gap between Paleozoic and Mesozoic: the lower half of the Karroo is Permian in age, while the upper half corresponds to roughly three-quarters of the Triassic. Were we to base our opinion strictly on the events in terrestrial life in the Southern Hemisphere, we would certainly not pronounce the Permo-Triassic boundary as the time and place of the greatest extinction in history. It was a relatively ho-hum affair down south.

Therapsid "reptiles," the famous "mammal-like reptiles," are the dominant animals of the Karroo fauna. The contrast is especially apparent in the upper (Triassic) portion: elsewhere, as we shall see, archosaurian reptiles, (dinosaurs and their forerunners) were running the show, and therapsids were far from dominant members of the fauna. But in Gondwana, there was a remarkable radiation of herbivorous and carnivorous therapsids and they had the lock on the niches of the terrestrial ecosystem.

I have already remarked that E. H. Colbert identified a labyrinthodont amphibian jaw from Triassic beds in Antarctica—and did so at a time when evidence in favor of drifting continents was building to such a crescendo that Colbert's discovery added force to the argument, rather than appearing as a *Mesosaurus*-like anomaly that need special explanation. Colbert had collected extensively from Gondwana deposits in South Africa, India, and South America. He knew the Antarctica deposits to be equivalent to the Upper Karroo. Colbert traveled to Antarctica with James Kitching, storied collector extraordinaire of South African Karroo beds who could tell from the helicopter thay they literally had outcrops of true Karroo beds below them; they managed to collect many specimens of Karroo therapsids that austral summer, dispelling any doubts whatever of the former connection between the continents.

In retrospect, it is easy to second-guess, to wonder why there was ever doubt at all on the former existence of the Gondwana supercontinent. As we have seen, it took geophysical evidence for all geologists to accept the reality of plate tectonics. Mariners had been obtaining coal to fuel their ships from local deposits associated with leaves of the extinct plant *Glossopteris*, already well known from deposits in India, South Africa, and South America, that had long been known from Antarctic deposits. But tradition dies hard, and such evidence in the prevailing climate of *stability* was simply brushed aside.

One final word before we enter the new world of the Mesozoic: the rich fossil record of therapsids in South Africa provides one of the very finest examples of anatomical evolution yet found in the fossil record. One difference between mammals and all other tetrapods, be they "reptiles," "amphibians," or birds, is the presence of three little bones in the mammalian middle ear. All the rest have but one: the "stapes." The stapes seems to have been derived from the bone that supported the jaw in various groups of fishes: the "hyomandibular." Evolutionists, naturally enough, wondered where the outer two bones added to the mammalian middle ear came from.

As so often is the case, evidence on a question came from more than one area of science. The three bones are the malleus ("hammer"), incus ("anvil"), and the stapes ("stirrup"). Embryologists in the latter part of the last century discovered that in early stages of embryonic development, the middle bone of the three (the incus) was positioned with tissues destined to become skull bones, while the outer bone, the malleus, seemed to come from the back region of the lower jaw. The sequence of therapsids in South Africa confirms the embryological investigations convincingly: in several lineages of these mammal-like creatures, two bones, the quadrate of the skull and the articular of the lower jaw, become progressively smaller and, in time, more closely associated with the ear region. When they finally disappeared as quadrate and articular, three bones have become established in the middle ear. The bones as quadrate and articular were always closely associated: in living reptiles, these are the bones that together form the joint between lower jaw and skull. In intermediate therapsids, there is a double articulation, between the quadrate and articular, in time-honored fashion of days of yore (and retained in birds, lizards, and snakes today) and between two other bones, the squamosal of the skull and the dentary of the lower jaw. Indeed, the dentary forms the entire jaw in mammals, whereas there are a number of bones in the primitive, "reptilian" lower jaw. Thus the fossil record provides a wonderfully detailed sequence of intermediate anatomical forms between the primitive condition and the derived, evolutionarily advanced state. It is a sequence that is progressive through the rock layers, and agrees perfectly with the sequence of events that occur in the development of the mammalian jaw through embryological history. It is striking cases such as these that abundantly confirm the

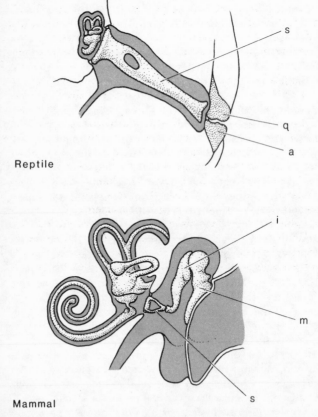

FIGURE 55 Diagram illustrating the transition of bones in the reptilian jaw joint to mammalian bones of the middle ear. Both reptile and mammal have a *stapes* (s), the "stirrup" of mammals. The *quadrate* bone (q) of the reptile becomes the *incus* (i) of the mammal, while the reptilian *articular* (a) becomes the mammalian *malleus* (m), also known as the "hammer."

strength and clarity of Darwin's original vision of the evolutionary history of life.

But the great episode of therapsid domination of the southern continents eventually came to an end. We have already seen how a naive expectation of progress in history is always sure to be dashed on the rocks of the actual events in life's evolutionary saga. The therapsids of the southern hemisphere are no exception. Their Waterloo came not at the end of the Paleozoic, nor even at the end of the Triassic, but during the later stages of Triassic times. In the upper beds of the Karroo sequence, in rocks known in South Africa as the Cave Sandstone, all manner of dinosaurs, already radiated into an array of ecological niches, abruptly show up. By Cave Sandstone times, the therapsids are already gone. It is impossible to say whether the dinosaurs swept in and took over, or the therapsids first suffered a severe extinction, vacating the territory and providing more opportunity for dinosaurs. My instincts favor the latter theory. But one thing is for sure: that steady, progressive climb toward mammal-ness that shines through the rock sequence of the Karroo did *not* culminate in the coronation of a new, superior bunch of hairy placental tetrapods—mammals—that inherited the Earth. That day was still some 145 million years away. It was the diapsids that were destined to control terrestrial ecosystems for the rest of the Mesozoic, while the mammals that had emerged by the end of the Triassic remained small, living furtively as minor denizens in a world dominated by dinosaurs. That the therapsids gave way to the dinosaurs shows that there is no inherent superiority of the synapsid line (which includes mammals) over the line that includes most of the familiar reptiles and the birds. Progress and superiority, so entrenched in our evolutionary vision, really have no place at all in an objective evaluation of the vicissitudes of life's history.

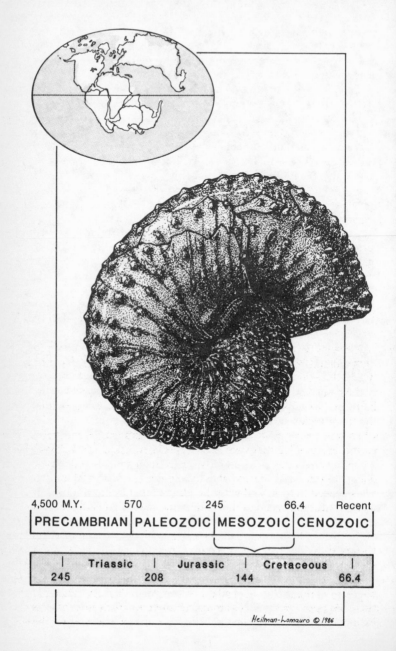

4,500 M.Y.   570   245   66.4   Recent

PRECAMBRIAN | **PALEOZOIC** | **MESOZOIC** | **CENOZOIC**

	Triassic		Jurassic		Cretaceous	
245		208		144		66.4

Heilman-Lomauro © 1986

6

Life's Middle Age: The Glorious Mesozoic

Dinosaurs are the quintessential fossils, the first and often the only creatures people think of when it comes to pondering life in the past. And no wonder: the Bible tells us of giants walking the Earth in days of yore, and the specter of huge creatures marauding the countryside has sent a chill up many a spine since the mid-1800s, when we first achieved a measure of understanding of these great, extinct beasts. That they were *reptiles* added to the fascination: that a group of vertebrates more primitive than the mammals which now dominate terrestrial life was able to rule the Earth for 150 million years seemed to fit in well with the idea of progress in evolution. It makes a neat tale: good as they were, dinosaurs finally relinquished their grasp of the Earth, yielding to the relentless and inevitable rise of the mammals, but, as we shall see, this is a tale based far more on wishful thinking than any objective assessment of the facts of the matter.

Nor is it only fourth graders entranced with dinosaurs who tend to see them to the exclusion of all other forms of extinct life. Recently there has been a tremendous upsurge of interest in the causes of mass extinctions—the most famous being the one at the end of the

Cretaceous, the event that ended the Mesozoic some 66.4 million years ago. That's when the dinosaurs became extinct once and for all—fantasies of them still living in the jungles of west-central Africa notwithstanding. Physicists and astronomers, as we shall see in Chapter 7, have been prominent in debating newer ideas about the causes of extinctions, and even *they* have a hard time seeing beyond the dinosaurs.

The simple truth is that many more creatures than dinosaurs were living in Mesozoic times. Life was as varied and rich on land and sea in Mesozoic times as it was in the Paleozoic, and as we see it around us today. Our goal here is to appreciate that variety—to strike a more balanced view of the fabric of Mesozoic life—before we tackle the perennially intriguing question of what happened to life (including the dinosaurs) when the Mesozoic finally did draw to a close.

The Middle Stage of Life's History

Mesozoic literally means "middle life." It is the middle of the three great chronologic eras of life on Earth. But beyond their chronologic position, there really is a sort of intermediate look to the forms of life preserved as fossils in Triassic, Jurassic, and Cretaceous rocks. This means that the gross forms of life as we know it today began to take on recognizable shape in their ancient relatives of the Mesozoic. Corals, for example, arose back in Cambrian times, but they were *rugose* and *tabulate* corals. Modern corals, as we saw in Chapter 4, are all scleractinians, with their aragonitic skeletons and six-fold symmetry. Rugosans and tabulates are the corals of the Paleozoic; they simply disappear from the record after the Permian, presumably numbering among the many victims of the mass extinction that drew the Paleozoic to a close. Scleractinians show up in Middle Triassic rocks as ecological replacements of the earlier corals.

It is the same way with group after group of living things: life in the Mesozoic becomes distinctly more modern than we were accustomed to see in the Paleozoic. The pattern has a direct effect on classification schemes: the Class Osteichthyes (bony fishes) arose in the Devonian, but the Teleostei (modern bony fishes) arose in the uppermost Jurassic or Lower Cretaceous—and today are nearly synonymous with the word "fish." Only a handful of more primitive types of osteichthyans are still living, while the number of teleost species in the world's lakes, streams, and oceans is in the several thousands.

But this gradual "modernization" process is but one theme of Mesozoic life. Life also had its surprises back then—utterly new sorts of ways to make a living, for example, undreamt of in the Paleozoic (such as the successful invasion of both air and sea by several independent lines of "reptiles"), and the exuberant exploitation of other modes of life that in our age seem forever lost. The group that offers the best example of the loss of an ancient heritage is the ammonites—the group that I and many fellow paleontologists think of *before* dinosaurs whenever anyone mentions the Mesozoic.

Ammonites

Amun was the chief god of the Triad of Thebes—the sacred threesome at the center of religious belief in that ancient city perched on the east bank of the Nile in central Egypt. When the political power of the local rulers grew strong enough to encompass all of "Upper and Lower Egypt," they became the pharaohs of the New Kingdom, and Amun became the chief god of Egypt.

Amun (spelled "Ammon" by the Greeks) was often depicted in the form of a ram, with large, incurved horns tucked in tightly at either side of his head. An "Ammonite" is a "stone of Amun"—a tightly coiled fossil shell bearing a superficial resemblance to a ram's horn. Actually, an ammonite is any member of a huge, long-ranging group of externally shelled cephalopod mollusks, met in passing when we encountered nautiloids in the Paleozoic (Chapter 4). Ammonites are in many ways the most important group of fossils from any time or place: with their vast history (arising, as discussed in Chapter 4, in the Middle Devonian and ending with extinction at the end of the Cretaceous—a group lifespan of at least 330 million years) and distribution around the world's oceans, they were often abundant and have left a dense fossil record. But it is the range of problems that paleontologists have tackled and solved by focusing on ammonites that is the most arresting aspect of these creatures. I discuss several of these below. In addition, ammonites have been very useful in correlating rocks of Upper Paleozoic and Mesozoic age—and hence are indispensable in deciphering Earth history throughout that large chunk of geologic time. Indeed, they were so useful that from the very beginning of the study of geologic time, ammonites have been at the center of our theoretical understanding of how to go about telling time with fossils.

Ammonite Beginnings

There is some dispute about ammonite origins. They first show up in rocks of late Lower Devonian age. Some paleontologists feel they were derived from bactritid nautiloids, which had straight shells, but with a slightly wavy suture and a siphuncle (connecting tube) located at the outer (lower) side of the skull—a feature of nearly all true ammonites. However, in the same rocks that have produced the earliest ammonites, there is a series of straight through partially coiled to entirely coiled nautiloids,, and some paleontologists have argued that ammonites sprang from a group of coiled-up, rather than straight, nautiloids.

It is the usual state of affairs in paleontology that good anatomical intermediates between larger groups of organisms are rare to nonexistent. Thus the earliest whales are true whales, albeit a bit primitive; the same is true of bats. We simply lack a good series of intermediates to show us from which sort of land-dwelling mammals the whales and bats (separately, of course!) arose. With the ammonites, we seem to have not one, but two pretty good candidates for intermediacy, a matter that will undoubtedly be resolved with further analysis. Ammonites arose fairly quickly, but by no means in any sudden leap—a good example of the derivation of one group from a branch of another which, like the therapsid reptile story and that of *Archaeopteryx* (the first bird), is an excellent rejoinder to the creationist claim that there are *no* intermediate fossil forms to be found between any of the major groups of organisms.

"Ammonites" really constitute the Subclass Ammonoida of the molluscan Class Cephalopoda. The usual story about the evolutionary history of the group—true as far as it goes—is that there are three major divisions of the Ammonoida: goniatites, ceratites, and ammonites proper. They correspond to the Orders Goniatitida, Ceratitida, and Ammonitida and together seem to form a wonderful example of progressive evolution: goniatites are exclusively Paleozoic, ceratites (a bit more complex) are exclusively creatures of the Triassic, and true ammonites (the most complex of all) typify Jurassic and Cretaceous times. Together, this three-fold division of "ammonites" (that is, ammonoids) supplied one of the best examples of evolutionary grades—broad groupings of organisms all living at roughly the same time and representing one coherent stage along a route of progressive evolutionary improvement. The concept of grades is closely tied in with the basic notions of adaptation and natural selection, still the twin cornerstones of modern evolutionary thinking; before I douse the ammonoid example with the cold water of reality, we should take a look at both the general concept of grades and some details of ammonoid history that really do seem to fit the notion rather well.

I have already noted that the notion of natural selection—that the relative success of organisms in the economic sector of life (making a living) has very much to do with how successful organisms tend to be in reproducing themselves—is so strong that it leaves most evolutionists with the very firm conviction that evolutionary change is simply inevitable given enough time. If we picture nature constantly undergoing change, and in particular natural selection constantly seeking to improve the adaptations of organisms, we can readily imagine that all manner of innovations, many of them destined to be short-lived, will crop up. Indeed, I mentioned in Chapter 4 that early ammonoid history bears the stamp of early "experimentation," in which a variety of designs was, so to speak, tried out before the basic ammonoid design for all times was settled upon. The clymeniids, considered an order all to themselves, lived in the Upper Devonian and became extinct "without issue," as the saying goes. Unlike all other ammonoids before or after them, clymeniids had their siphuncle (that tube connecting the inner chambers) at the top rather than the bottom of the shell. There were other differences as well; yet clymeniids sported the standard "goniatitic" suture pattern (which I will discuss in a moment). Thus they sprang from regular ammonites; they were really just a sidetrack, and it arguably makes some sense to see them as "experimental" forms. Nor is their extinction without leaving any detectable descendant trace necessarily to be construed as a sort of evolutionary "failure." When extinction claimed them, as it did so many different groups at the end of the Devonian, the situation could well have been the other way around: they might well have made it through, and the goniatites might have failed. The rest of ammonite history would have been, presumably, quite different from what we are about to see unfold. Once again, as we saw with "living fossils" (Chapter 4), the survival of one group and the demise of a close relative seems not to be a simple function of some imagined "superiority" of one group over another.

The concept of experimentation leads, further, to the notion of parallel evolution: that a certain level of complexity of structure may be attained independently by several related lineages. We encountered this notion with the arthropods, particularly in their multiple independent invasions of the land. Particularly when such advance is achieved at more or less the same time, we recognize a new "level of biological organization," or, in other words, a grade. Members of a grade may all be descendants of the same common ancestral species, but in the heyday of the concept's popularity (mid-1950s to mid-1960s), the emphasis was on the supposed attainment of some level of evolutionary advance more or less at the same time by a number of different lineages—strong testimony, indeed, to the power of natural selection to effect general improvement in a number of separate lineages within a group.

How well, exactly, do the facts of ammonoid history fit this concept of grades? Over the vast history of ammonoids, the shells of different lineages of ammonoids took on a vast array of shapes and sported a number of different ornamental patterns: some were huge, like the five-foot-diameter shells of *Placenticeras* from the Cretaceous of the American West. (The French Lieutenant, whose woman furnished the title for John Fowles's novel, was an amateur paleontologist who collected ammonites in the Jurassic cliffs at Lyme Regis, a town on the Dorset coast of southern England. Fowles himself is the unpaid curator of the local paleontological museum. The ammonites there are justly renowned, some reaching diameters of two or even three feet.)

But the theme of grades is based on the suture patterns formed when the internal partitions intersect the outer walls of the shell. (Recall the details of shelled cephalopod anatomy from Chapter 4, p113.) Goniatite sutures are rather simple, wavy affairs, with but a few major "lobes" and "saddles." "Saddles" are deflections of the suture directed toward the end, or "living chamber," of the shell, while "lobes" point backward. Unlike goniatites, the suture patterns of ceratites and ammonites are far more involved: ceratite lobes and ammonite lobes and saddles are complexly crinkled. Most Paleozoic ammonoids have goniatitic sutures; nearly all Triassic ammonoids have ceratitic sutures—and virtually all Jurassic and Cretaceous ammonites have ammonitic sutures. But there are exceptions: there are some Mesozoic genera with sutures that are rather simple and goniatite-like, and some post-Triassic genera with rather ceratitic-looking sutures. The reverse is also true: some Upper Paleozoic ammonoids have distinctly ceratitic and even ammonitic-looking sutures. According to most ammonoid specialists, though, these anomalies are *convergences*: independently derived similarities. Thus the consensus is that, for example, those Cretaceous forms that have goniatite-like sutures are actually most closely related to other Cretaceous genera with true ammonite-like sutures—rather than being holdovers from the Paleozoic. The analysis, in each case, is based on similarities in other parts of the shell that link the anomalous genera up with other "normal" genera living at the same general time.

But overall the evidence is striking: almost any ammonite you are ever likely to see will follow the rule, so much so that one can reliably reverse the logical sequence and use the method to provide a gross dating of rocks. You are almost certainly in the Triassic if you find an ammonoid with a ceratitic suture, and so forth. Ammonoid evolution was actually so rapid that in general you can do a lot better than that: rocks are datable in increments as brief as a million years on a

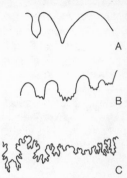

FIGURE 56 Diagram comparing goniatite (A), ceratite (B), and ammonite (C) forms of ammonoid suture patterns. The goniatite *Imitoceras* is shown with sutures in place on the shell, and additionally with the suture pattern drawn flat on the planar surface of the page. This latter convention, common among ammonoid paleontologists, allows ready comparison with other patterns, as with the ceratite and ammonite patterns.

worldwide basis in much of the Mesozoic—based strictly on an analysis of the ammonites in a given sample.

Ammonites and "Racial Senescence"

Ammonites have also played a pivotal role in our grasp of the very nature of the evolutionary process. They have been used as examples of a wide variety of evolutionary phenomena—some rather fanciful, others much more solidly based in fact. An imaginative but spectacularly wrong use of ammonites in evolutionary thinking can be found at the close of ammonite history. There, late in the Cretaceous, we begin to find a greater than usual incidence of so-called *heteromorphs*, a name alluding to the rather unusual (not to say downright weird) shell forms of some species of ammonites. Departing from the standard coiled shape, some shells were bizarrely twisted into something resembling a knot, while others were looped around as J-shaped hooks. Still others, like the common genus *Baculites*, began life as a normal spiral, then grew straight as an arrow, resembling their ancient, straight-shelled nautiloid relatives. Seventy-five years or so ago it was fashionable to see the heteromorphs as deviants from the norm, and thus as evidence of the supposed phenomenon of "racial senescence." Extinction of the ammonites was drawing near, and, it was supposed, the lineage had run out of evolutionary steam;

old, tried-and-true genetic laws of regular growth were breaking down, much as the cells in the body of an aging organism lose their vigor as death approaches. Extinction seemed partly, at least, a matter of internal decay, built into the system of species and larger groups much as aging and death are built into the makeup of individual organisms.

FIGURE 57 A scaphite—a Cretaceous heteromorphic ammonite of the sort that formerly inspired the notion of "racial senescence."

No idea in science is inherently bad: ideas are just ideas, and in the case of "racial senescence," at least its sponsors were making what I think to be a profound, and too often ignored, analogy between organisms and larger-scale entities, which are also "individuals" in their own right. Species and larger groups of organisms (such as the entire Order Ammonitida of the Class Cephalopoda, Phylum Mollusca) definitely do have births, histories, and deaths in their own right—just as organisms do. So it makes sense to wonder to what degree the deaths of species are analogous to the deaths of organisms, and, in particular, to ask whether there is anything internal, within species, that causes in whole or in part their eventual demise.

But if ideas cannot be inherently bad, they can quickly become so as soon as we bring in that *other* ingredient of science: the real world. It is the aim of science to provide an accurate description of the material universe, a description so general, and so complete, that we gain an understanding of what things exist and the rules underlying their appearance and disappearance. That, in a nutshell, is what science is all about. If there is any one cardinal, procedural rule in science it is simply that scientific ideas—if we are to accept them, however provisionally, as a valid description of the way things really are—must agree with what scientists actually observe in nature. To test an idea, commonly we make predictions: *if*, we say, this idea is true, then nature should show us certain consequences.

Extinctions have prompted the greatest collection of bad ideas in the entire history of paleontology. I'll discuss a number of them in the next chapter, but I'll deal with racial senescence here, as it directly involves the ammonites. If it were true that ammonites died off as a collective whole because they became unable to obey the general laws of growth they had always followed in the past, or because they had evolved to such an outlandish anatomical degree that they couldn't hope to function very well, we would expect that all ammonites at the end of the Cretaceous would be heteromorphic—that is, oddly coiled. Moreover, we would predict that ammonites in "normal" times—i.e., when extinction did not lay just ahead—never assumed such unusual shapes. We would be wrong on both counts: in a fashion utterly typical of life's overall history, there were present at the end of ammonite times both heteromorphs and perfectly normally coiled shells. Actually most heteromorphs lived in the Lower Cretaceous, and were pretty much gone long before the famous extinction event occurred. And heteromorphs dot the ammonite landscape here and there at other times in their long history—definitely not presaging any major extinction episode (though admittedly there were some heteromorphs in the Upper Triassic, and there *was* an important ammonite extinction event at the end of the Triassic). Nor would a desire to fashion an internal reason for ammonite extinction survive a reckoning with the most obvious phenomenon of all: that great terminal Cretaceous extinction that cut across genealogical lines, taking out algae, reptiles, and fish as well as ammonites from the seas, and taking out a vast range of land-living creatures as well. If it is a mistake to think only of dinosaurs when considering that extinction, it is equally erroneous to think only of ammonites.

Thus racial senescence has been decisively discarded as an explanation of extinction. The great vertebrate paleontologist George Gaylord Simpson gave the *coup de grace* to this and similar ideas that rely on internal factors as major causes of evolutionary patterns. Racial senescence was invoked for other famous examples as well, such as the "Irish Elk," a huge deer of the European Ice Age (Pleistocene) whose antlers were so large that many people openly speculated that the creatures could barely walk. A sure example, they thought, of nature going too far, producing structures that were actually detrimental to the organisms, thus contributing factors to their extinction. Yet, it is clear that the Irish elk enjoyed a long history, the antlers were perfectly functional (used presumably in courtship displays and battles, just as modern deer utilize their antlers). Once again, their extinction was tied into general, across-the-genealogical-board extinction of many species of large mammals as the last glaciers retreated to the north. Still other examples were simply based on false observation: there is a famous coiled oyster, *Gryphaea*, of the Mesozoic that, it was said, became so incurved that the small upper shell that functioned as a flap-like covering over the soft animal ensconced in the bigger, coiled shell below, simply could not be opened. X-rays have shown the story to be untrue: what looked like shell covering the upper flap and forcing it to remain forever shut was simply hardened mud that had not been properly cleaned away when the specimen was studied. Simpson argued that all such cases were inherently impossible anyway: natural selection, working to shape and maintain adaptations, could hardly be expected to fashion features actually harmful to organisms. Nor was there any scientific evidence then (or now) that would lead us to

FIGURE 58 Reconstruction of the extinct Pleistocene "Irish elk"—another supposed victim of racial senescence.

accept some internal factor within organisms that would force them to keep evolving in some direction to such an extent that they would eventually become extinct. The issue of "racial senescence" remains a closed subject, but one that could potentially be resurrected if, and only if, some genetic mechanism emerges from molecular biology labs that would supply us with plausible scenario of how it might occur.

"Ontogeny Recapitulates Phylogeny"

Ernst Haeckel was one of Darwin's most enthusiastic supporters on the European mainland. Soon after the *Origin of Species* appeared, Haeckel was off and running in the new-found game of reconstructing the evolutionary history of life. To some extent the job was already well underway, for there is a pattern of similarity linking up all organisms, both living and extinct. Darwin simply showed that pattern to be the natural byproduct of the evolutionary process—thus many of the older classifications could readily be interpreted as representations of evolutionary, or *phylogenetic*, affinity. But it was Haeckel, as much as anyone else, who got into the business of drawing "phylogenetic trees" depicting in detail the reconstructed patterns of the history of life. And along the way Haeckel formulated some general rules of procedure, based on his analysis of the typical patterns of evolutionary history. Easily his most famous—still learned in high school biology classes—is the so-called "Biogenetic Law" or the "Law of Recapitulation": *ontogeny recapitulates phylogeny*. According to Haeckel, the development of an organism in some sense goes through the general stages of the evolutionary development of its entire lineage. Through much of the last 150 years of paleontological research, ammonites have been considered to be prime examples of Haeckel's dictum, which carries with it the promise that one can reconstruct the evolutionary history of an entire group simply by examining the developmental history of a single organism.

The history of Haeckel's idea, based on the older notions of the embryologist Karl von Baer—is almost comical. Once pronounced the wisdom of the ages, the law has been subjected to a great deal of ridicule in the past fifty years. And taken literally, as some less imaginative biologists and paleontologists actually did take it, the idea is indeed absurd: no one supposed that the stage of human (or dogs, for that matter) development, where slits in what is to become the "neck" are apparent, literally represents what our ancestors looked like when they still had functional gills in the adult stage. At that point in vertebrate history, of course, those ancestors looked like fish—because that's what they were. Embryos are precisely that: embryos. They do not represent what ancestors looked like in complete adult form. But they do often show characteristics of those ancient adults. I have already remarked that those slits in vertebrate embryos do not have gills in them; but they still are holdovers from an era when all vertebrates had gill slits lined with functional oxygen-deriving organs—gills—in the adult stage. And that's where von Baer and Haeckel come in: ontogeny does not recapitulate phylogeny in any strict, literal sense. But in looking at early stages of development, we often see characteristics later lost in development, characteristics that were present in the ancestor. It is a handy way of assessing evolutionary affinity—just as Haeckel said it was.

Now, ammonites are mollusks; most mollusks have shells, and those shells are ever-growing skeletons. In a very real sense, in the form of their shells, mollusks drag their developmental histories around with them. Ammonoids are no exception, and since they, like nautiloids, periodically lay down a new partition, each shell can be peeled back to examine progressively earlier and earlier stages of development of the shape of that all-important suture line. In goniatites, things are always fairly simple; in the Carboniferous, though, goniatites were adding a few lobes and saddles as they got bigger. By the time we get to the ceratites and the even later ammonites, we see many more lobes and saddles, plus more and more crinkling, added to the shell as the animal kept getting bigger.

If the course of evolution was heading toward bigger shells in a lineage, or simply toward the increased complexity of the suture lines, new patterns would tend to be added on to the end of the developmental scheme. Such evolutionary modification, by simple terminal addition, would automatically allow us to look at earlier stages of development of some advanced species and deduce, pretty accurately, what form of suture was characteristic of its ancestor. We would then look for that predicted suture pattern—and it is a pretty common experience in ammonoid paleontology that the predicted patterns are found in ammonoids collected from older rocks. We might also expect that as evolution proceeds, there would be a need to drop out some of the earlier stages, there being not enough room on the shell (or time in the ammonoid's life) to go through all the ancestral stages. And that is also what is found, at least according to some paleontologists.

That being said, I will now acknowledge that the usual cry against Haeckelian recapitulation also maintains that evolution need not proceed by simple addition to the end of the developmental process;

indeed, the condition variously known as neoteny or pedomorphosis (the two are technically distinguishable as variants of a theme) proceeds in many respects in just the opposite manner: new organisms appear in evolution because larval forms become sexually mature and stop further development, or larval characters are retained well into the normal stage of sexual maturity. The standard example is the axolotl, a Mexican salamander that keeps its gills and retains a tadpole's existence throughout its life, as a young, true tadpole and as a sexually mature adult. That this can happen is well extablished, and, of course, an exception that sorely tries Haeckel's "Law."

But acceleration, as the usual mode of evolutionary transformation of the ammonoid shell is technically termed, is very much the rule in ammonoids—a useful feature, indeed, when it comes to interpreting the actual pathways of ammonoid evolution. And this rule holds so well with ammonoids in general that it seems to reinforce the general notion of grades through ammonoid history: suture patterns become more complex as the animal grows, and the increase in complexity does seem to have a relationship to the course of events within each ammonoid lineage. Does this not fit in rather well with the general course of ammonoid history—from Paleozoic goniatite to Triassic ceratite to, ultimately, Jurassic and Cretaceous ammonite? Superficially, yes. Actually, no.

Grades, Sutures, and Ammonoid History

Figure 59 is a sketch of ammonoid history. It is one of the most instructive diagrams I have ever seen. A few points immediately leap out, enough to gladden the heart of anyone who would really like to see what the fossil record has to offer us in the way of a deeper understanding of the evolutionary process.

First of all, we indeed see the Goniatitida, Ceratitida, and Ammonitida arrayed as advertised in geologic time and, for the most part, totally dominating ammonoid diversity in their respective time slots. But we see other orders as well: the Anarcestida, Prolecanitida, and the aforementioned Clymeniida of the Paleozoic, for starters. All nonetheless conform to the general goniatite mode of sutural development, so that the gradal story remains intact at this level. But look at the ceratites in the Triassic: ammonoid experts agree that they are a unified ("monophyletic") group, all right. But they were seemingly not derived from the Goniatitida, but rather from the far less diverse and rather conservative lineage, the Prolecanitida, which just happened to send a few genera, more or less unchanged,

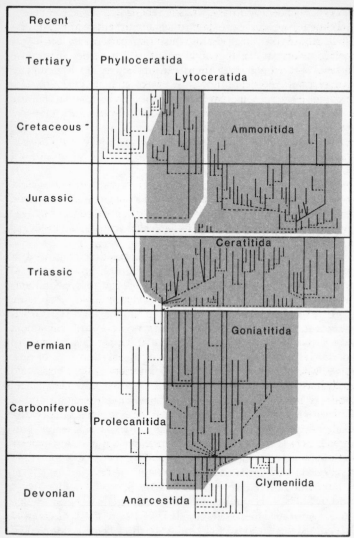

FIGURE 59 The ammonoid family tree.

across the great divide of the Permo-Triassic boundary. No pattern here of independent development of the ceratitic suture, and certainly no direct evolution from Goniatitida to Ceratitida.

And the pattern is once more repeated: ceratites, a coherent evolutionary group, diversify all over the Triassic seascape. Just as

phylogenetic history is far from gradual, its pattern hardly a documentation of successive, progressive "grades" of improvement. Its history is one of great stability, disrupted by greatly spaced catastrophes (which ammonites shared with many other groups) and subsequent reproliferation based on what little genetic information managed to survive. The ammonites give us an enthralling picture of the gross patterns of the pulse of life—and a picture rather drastically different from conventional, text-book imageries of evolution.

Life on the Rebound

Ammonoids bounced back after their flirtation with catastrophe at the close of the Paleozoic. The theme is an old one: after the entire lineage coming ever so close to annihilation, just a little bit of genetic information manages to survive to provide the raw material for a renewed phase of evolutionary activity. For that information retains the instructions on how to make baby ammonoids and how those ammonoids will grow to be big ammonoids—and eventually make still more baby ammonoids. Not all the various forms of ammonoid genetic information made it into the Mesozoic: the extinction greatly biased which particular set of ammonoid genetic instructions was to form the basis for future ammonoid evolution. But granted that *any* of that information made it through, it forms the basis for more mutations, natural selection, and additional evolutionary processes (such as speciation) to modify and diversify the ways of being an ammonoid. History, especially the evolutionary histories of organisms, will not be repeated in any exact sense—too much of the old information is gone. But we are likely to see a repetition of old adaptive themes—ways of being an ammonoid that worked in the past and are refashioned when life starts bouncing back.

And that really is the theme for all of early Triassic life. Few kinds of organisms are evident in the rock layers immediately above the extinction. But after a few milion years, there is always a sudden proliferation, very much as if life has exploded in all directions. Old niches are filled with new faces, and the pattern is as evident on land as it is on sea.

But the newness of the creatures filling up habitats left vacant by these massive extinctions may lead us to overlook those creatures that seem to have made it through relatively unscathed. As we shall soon see, when it comes to pondering the causes of extinctions, the question must be asked: why did this group make it through, but not that one? Why, for example, did dinosaurs succumb at the end of the Cretaceous—while, for example, crocodilians sailed right through?

Such survivorship is important, too, in our consideration of what happens *after* the episodes that remove so many different kinds of animals and plants. Paleontologist Roger Batten of the American Museum of Natural History has pointed out that many of the archaic forms of gastropods (snails) known from the Paleozoic did, indeed, make it into the Triassic. Curiously enough, such Paleozoic sorts of snails as the bellerophontids are never found right at the base of the Triassic: it is usually well into Middle Triassic times before we start to pick up these odd-ball relics of the Carboniferous and Permian in our samples. Nor is there any realistic possibility that these primitive snails were re-evolved: Middle Triassic specimens seem quite similar to the older models.

Thus there is typically a hiatus, a large gap in time in the distribution of some organisms. And these gaps are often associated with mass extinctions. The answer to the puzzle, or so Batten and David Jablonski of the University of Chicago have speculated, is that there are *refugia*, isolated regions of the world's oceanic systems in the case of marine mollusks, where some species manage to survive. They are highly localized, and have left little or no fossil record behind them in the regions to which they became restricted as extinction claimed their relatives elsewhere around the world. Indeed, if they were living around oceanic islands such as those we now see in the South Pacific, as Jablonski has claimed, then there is every chance that the old habitats were destroyed as the vast crustal plates were consumed at their leading edges. Even had they left a fossil record, it would have been destroyed long ago.

Terrestrial Life in the Early Mesozoic

Life on land follows the same general ecological and evolutionary rules, and so leaves much the same sort of patterns we have seen time after time in the sea. Therapsid reptiles were bloodied but pretty much unbowed at the Permo-Triassic boundary, and particularly in Gondwana they seem not to have been too drastically affected by the events that wrought such havoc elsewhere around the world. Nonetheless, by the end of Triassic times, the complexion of life on land had also changed—life had entered its "middle age" there just as it had in the sea. And what better symbol of changing times could there be than the advent of the dinosaurs?

Paleontologist Mike Benton of Belfast, Northern Ireland, has recently called the Triassic the "Age of the Rhynchosaur." Although it is a somewhat tongue-in-cheek suggestion (few people, after all, have ever heard of rhynchosaurs!), Benton nevertheless has a point: rhynchosaurs were among the most common Triassic land animals. Like dinosaurs, they belong to that other great division of amniote life: the Diapsida (the reptiles with two holes on each side of the skull). As Benton makes plain, it is not that rhynchosaurs were as diversified as, say, therapsids or, for that matter, dinosaurs themselves: all of them seem to have been rather lumbering herbivores, and there seems not to have been a carnivore in the bunch.

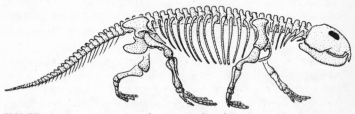

FIGURE 60   A reconstruction of a Triassic rhynchosaur.

Yet there were no true dinosaurs until *after* the rhynchosaurs became extinct. Rhynchosaurs survived well into the Upper Triassic. Then, just a bit later, we find early dinosaurs (forerunners of the titanic *Brontosaurus* and *Diplodocus*) occupying the herbivore niche; rhynchosaurs are by then utterly absent—missing and presumed extinct. It certainly looks as though dinosaurs took over a niche that had been for a goodly measure of time (most of the Triassic—some 35 million years) and to a great degree the province of the rhynchosaurs.

And Benton asks us to ponder this pattern of replacement. Were the dinosaurs simply superior to the rhynchosaurs, driving them out of their accustomed haunts and habitats? Or, as Benton prefers, was it simply a matter of rhynchosaurs running into ecological trouble, finally yielding to extinction, thus vacating a series of niches that other groups—notably the ancestors of the dinosaurs—were only too happy to take over, there to radiate themselves?

Benton makes a good case for the latter argument. He ties the rhynchosaurian decline and fall not to the rise of dinosaurs, but to the loss of their favored foodstuff, which he believes was the seed fern *Dicroidium*. (Interestingly, and a bit ironically, Benton sees the extinction of *Dicroidium* as the result of direct competition with the spread of true conifers. Thus he prefers not to think in terms of competition inducing extinction when it comes to the interpretation of the evolutionary histories of animals, but with plants the old notion of superiority of one "advanced" form over a primitive forerunner is apparently still acceptable!) Thus, to Benton, dinosaurs radiated partly as simple opportunists—and once again we seem to have a situation where nature abhors a vacuum: extinction simply opens up vacancies, creating ecological opportunities bound to be filled immediately. If we ask how dinosaurs managed to dominate mammals throughout the Mesozoic, and if we see mammals not so much as superior creatures that finally got the best of dinosaurs but as humble, modest little things that got nowhere in an evolutionary sense for the 130 million years or so that the dinosaurs were in control of terrestrial habitats, we come up with exactly the same conclusions: mammals diversified only after the dinosaurs had gone. The meek—or the just plain lucky—inherited the Earth. Benton simply wants to read the same moral into the rise of the dinosaurs themselves. And his picture is, in many ways, compelling.

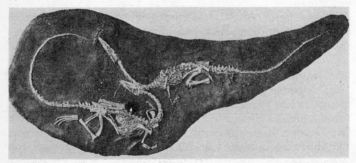

FIGURE 61 Two specimens of the small early dinosaur *Coelophysis* from New Mexico.

The Dinosaur Industry

You can buy a full-scale, bone-by-bone replica of a *Triceratops* skeleton from the American Museum of Natural History for approximately $13,500 (as of this writing). This is not an ad, but an economic fact: dinosaurs are prohibitively expensive these days—and if it is not exactly true that everyone wants one, enough museums do to persuade large institutions, with good collections, to furnish scientifically accurate reproductions of existing specimens.

It is not that we have run out of dinosaurs in the fossil record, creating a scarcity that, according to the laws of supply and demand, will drive up the cost of obtaining one. Dinosaur aficionado Jim Jensen is rumored to know the locations of tens of good-quality skeletons jutting from the ground of Wyoming, Colorado, and Utah—where the Jurassic Morrison Formation is still an incredibly rich source of fossils of a wide variety of dinosaurs (and the bones and teeth of those little mammals running around the interstices of the dinosaurian world). No, it is not lack of specimens: it is simply the time and effort—the *money*—needed to dig those specimens out, transport them safely to a university or a museum, have them cleaned, and, ultimately, have them mounted that makes the collection of a new dinosaur specimen a rather rare event these days. It is far easier to order one from a catalogue.

Dinosaurs really did rule the Earth: we associate "rule" with brute force, and naturally think of those huge herbivores, such as *Brontosaurus*, and the fearsome gigantic carnivores, like *Tyrannosaurus rex*, which of course epitomize all that is exotic and frightening about the world of the past. But I mean something a bit different when I say "rule" here: for there was a vast array of creatures filling all sorts of ecological situations. There were carnivores of all sizes, including some little chicken-sized species that preyed on smaller items of the food chain. There were herbivores, and, as is usual in Nature, these came in an even greater array of shapes and sizes than carnivores, and belonged, as well, to a number of different groups. Thus ceratopsian dinosaurs (of which *Triceratops* was among the last), the ankylosaurs (which looked like armored tanks), the stegosaurs (with their double row of plates running along the back), and the great array of ornithopods (duckbills, hadrosaurs, and the like) were four different lineages comprising the herbivorous group Ornithischia. The sauropods (the *Brontosaurus*-like dinosaurs) were yet another branch of herbivores—closely allied with the meat-eating theropods (*Tyrannosaurus* and kin) and, as latest thinking would have it, with birds.

But dinosaurs were really *common* too—another aspect of "ruling the Earth." They were the standard-issue creatures of the day, and individual dinosaurs were as numerous as mammals are today. Anywhere you go where there are sedimentary rocks of appropriate age (Upper Triassic-Upper Cretaceous) and environment (meaning mostly lake and river deposits on continental surfaces), you are likely to run across traces of their former presence. In the great Upper Triassic-Lower Jurassic intermontane basins that run along the Atlantic coast of the United States from the valley of the Connecticut River down (sporadically) into North Carolina, fish, plants, and invertebrates come from the black shales of the ancient lakes, while the reddish beds (used to make all those New York City brownstones at the turn of the century) turn up three-toed footprints of the dinosaurs themselves, as well as the occasional bone.

FIGURE 62 Symbol of life's past—or at least of its more terrifying aspects: the "King Tyrant Lizard," *Tyrannosaurus rex* from the Upper Cretaceous of Hell Creek, Montana.

Cope and Marsh: Of Rivalry and Finite Paleontological Riches

Coastal plain sediments in New Jersey, and down in the Gulf Coast states, have produced a number of Cretaceous dinosaur bones—washed in by rivers emptying out in the sea. But it is the West that has produced the vast majority of North American dinosaur fossils—and some of the most colorful moments in the annals of paleontology. The fossil-bearing rocks stretch on endlessly, yet the opportunities to collect a wide variety of fossils are, as we shall see in a moment, somewhat restricted. The latter half of the nineteenth century saw rival bands of fossil hunters vying for specimens—sometimes violently. There were hostile bands of native North Americans with which to contend, as well as the more usual perils of collecting: hunger and thirst, terrible weather, accident and injury as cliffs were scaled and specimens pried loose from their rocky matrices. The early days of fossil hunting on the high plains are inextricably bound up with all the romantic imagery we still associate with the American West.

Nor is it all exaggeration. Part of this competitive frenetic exploration and collecting arose from a motive surprisingly similar to the force that impelled so many to "go West": opportunity. Sometimes it was simply a desire to start afresh; pure greed also prompted many a move from East to West. In the case of the fossil hunters, the riches were bones, not gold and silver. But the stakes were taken every bit as seriously, and therein lies some insight into the scientific mind. If scientists turn out to be as competitively venal as any other segment of mankind, it may be that the energies they expend on trying to get ahead, to best their rivals, will do some good after all: humanity is likely to be enriched in knowledge, and remain unaffected by the less attractive behaviors of the scientists themselves. Nowhere do we see this better than in the adventures of the two bitter rivals, the nineteenth century paleontologists Edward Drinker Cope and Othniel Charles Marsh.

Rivals are as common in science as in any other walk of human life. Roderick Impy Murchison, who traced the Nile to its sources, fell out with his friend Adam Sedgwick. Sedgwick described a sequence of rocks in Wales and western England that he called the "Cambrian." Murchison described another, younger sequence nearby—his "Silurian." The problem: rocks in the upper part of Sedgwick's sequence overlapped the older rocks in Murchison's system, and each, naturally, claimed the overlap as part of *his* rocks. It was not until Charles Lapworth, after Sedgwick and Murchison were dead, named the overlap the "Ordovician" that the controversy was resolved—in a simple act of compromise that would have pleased neither of the combatants.

Cope and Marsh heard reports of vast vertebrate paleontological riches—where bones were commonly seen weathering out of the badland topography that formed much of the landscape between the

Mississippi River and the Front Range of the Rocky Mountains. They had seen Joseph Leidy's early publications, and decided to join in on the search. There seemed plenty enough to go around, and at first (in the 1870s), there was little in the way of rivalry between the two.

Tradition has it that Cope was a vain, neurotic genius, a man who was driven and who could not maintain any financial stability. Marsh, on the other hand, is usually depicted as rich, although not as bright as Cope, thus unable fully to take advantage of his more comfortable circumstances. There is some truth to these images: Cope did have problems, both financial and otherwise. He spent the greater part of his career at the University of Pennsylvania in Philadelphia—and had to sell his specimens to finance his further operations on more than one occasion. Indeed, Cope's dinosaurs form the nucleus of the collections at the American Museum of Natural History, whereas Marsh's material remains at Yale, where Marsh was long ensconced at the Yale Peabody Museum.

Yet both men were indefatigable and gifted workers, both in the lab and out in the field. Their enmity arose, it is usually said, because even a space so vast as the great American West was not large enough for their two colossal egos. But the real reason is more human-scale. For though the space is vast out there—indeed, before the advent of the internal combustion engine, the space must have felt virtually limitless—and although the fossil-bearing strata stretch on for literally miles, there is nonetheless a peculiarly finite array of fossils that comes out of this treasure trove. And the finite nature of the array was bound to bring these two giants into competition.

This peculiarity in the fossil record is something I have yet to see adequately addressed by anyone. You can go to an outcrop—in the Upper Ordovician, say, just outside of Madison, Indiana—and pick up literally millions of fossils. Most of them will be so well preserved that it will not be difficult to sort them into separate piles of very similar specimens. And even if you know in your heart that only a few of the many specimens that ever lived actually contributed a shell to each of your piles of sorted specimens, you cannot help but feel that the fossil record is rich indeed.

Yet you may have, after exhaustive collecting, perhaps fifty or sixty piles—collections of specimens of different "kinds" of organisms, each pile probably corresponding to a different biological species. You have every reason to suppose that you have a sample of almost every hard-shelled invertebrate species that lived there in Upper Ordovician times—if not quite all species, at least *most* species. And you would probably be correct. But then you travel, say, to the Upper Ordovician of adjacent regions in Ohio, and you collect there as well. Soon you begin to notice that you are getting very much the same sort of things that you had been finding in Indiana.

And that is what I mean about the strange aspect of the fossil record: those formations that tend to produce fossils in abundance often are distributed rather widely. Organisms also tend to occur rather widely—in fact, organisms will occur wherever favorable habitats occur, so long as those habitats are in the same basic neck of the woods, such as the western half of a continent. What this boils down to is simple enough: with many exceptions, it is still generally true that fossils common in any one place will be found elsewhere. What had seemed like inexhaustible riches turns out, on closer examination, to be a repetitive sequence of similar fossils turning up over and over again all across a broad region. The number of *fossils* might seem limitless, but the number of *different species* you are likely to find is often quite limited.

Cope and Marsh worked in an age when discovery was all. Their main objective was to find and *name* as many fossils as possible that had never before been seen. In their zeal, and as soon as they found out that they had inadvertently given different names to the same species, the intensity of their competition picked up: there were even instances of telegraphing descriptions of fossils to the *New York Times*. Their jealousy seems to have prompted a sort of "dirty tricks" campaign, where shipments of fossils were diverted—or even, apparently, appropriated.

But juicy as many of those anecdotes are, it is what the two men, and their colleagues, actually accomplished that I find the most arresting. A tome 1,009 pages long and still affectionately called "Cope's Bible" consists solely of his scientific description of Tertiary-aged vertebrates (mostly mammals, but including fishes, amphibians, and reptiles). And those guys were good: haste does indeed make waste, but usually when lesser talents are involved. In general, their work has stood the test of time. Rough and ready as they seem to have been, Cope and Marsh were vibrant intellects whose work laid the very foundations of American vertebrate paleontology, including the study of dinosaurs.

One place that attracted both men was Como Bluff, in eastern Wyoming. Bone Cabin Quarry, still an active site of excavation, was named for a modest little hut built by a sheep herder. There being relatively few trees in the vicinity, the man fashioned his cabin out of the next best thing—the long leg bones of gigantic dinosaurs. Occasionally you really do stumble onto some amazing, untapped virgin fossil territory.

The bones of Bone Cabin and its quarry are all derived from the Morrison Formation, a geological unit that swings southwards down into Colorado and eastern Utah. Indeed, it is the very same formation that is featured at Dinosaur National Monument at Vernal, Utah—where half-excavated bones are beautifully displayed still lying in place in the wall of a steeply-sloping cliff face. And here is just the point about far-ranging species, dinosaurian or otherwise: most of the species at Vernal (first excavated by Earl Douglas of the Carnegie Museum around the turn of the century) also occur way over at Como Bluff, some 200 miles to the northeast. Many of the skeletons that have yet to be collected belong to well-known species:

FIGURE 63   Excavating a large dinosaur at Bone Cabin Quarry on Como Bluff in 1898.

FIGURE 64 The famous paleontologist Barnum Brown, discoverer of *Tyrannosaurus rex*, in the field at Sweetwater, Montana—fur coat and all.

today, as in the 1880s, paleontologists want new blood, specimens from species never found before—specimens they too can describe for the first time in the scientific literature, and to which they can give new names. It was relatively easy in the days of Cope and Marsh; but all the good fields have by now been picked over pretty thoroughly. Most of the basic descriptive work is done. If things got tight for Cope and Marsh—in terms of finding a constant supply of new creatures to name—imagine what it is like now. It really is a fairly rare phenomenon for someone to come up with a totally "new" fossil these days—though it does happen.

What then is the answer? Do we abandon the field because the early birds got all the worms? No, we don't—we change the game a bit, and find that there is still plenty to do. Dinosaurs are a wonderful case in point. Instead of spending all their time describing specimens, modern students of dinosaurs are busy trying to figure out just what on earth dinosaurs *were*: what their lives were like, and just how they evolved, which is to say, what they were and *are* related to.

Dinosaur Warmth

Just about everyone these days is saying that dinosaurs have had a bum rap: they certainly are not inferior to mammals. Just because they rose to prominence first, and, after a 150 million year run, fell victim to extinction is by no means a reason to relegate them to some lower rung of the evolutionary success ladder. But dinosaur enthusiasts are busy saying more, *much* more, about how remarkably good those creatures really were. The emphasis these days is on dinosaurs as efficient machines for living.

There is an old, and none too accurate, distinction between "warm-bloodedness" and "cold-bloodedness." The main issue has to do with maintenance of internal body temperatures. Birds and mammals (independently, as anatomical detail graphically reveals) have evolved four-chambered hearts. They have also developed feathers and hair, respectively, from the more primitive scales of reptiles—structures that greatly aid in *thermoregulation*. Together with additional physiological structures and mechanisms, birds and mammals have acquired the ability to maintain a constant body temperature to a great degree.

FIGURE 65 Alley Oop lives: skeletons of *Brontosaurus* and *Homo* side by side. Humans share general mammalian physiological mechanisms to regulate body heat; large dinosaurs apparently could maintain a constant body temperature by dint of their vast size alone.

Not so the reptiles. Most lizards and snakes, for example, experience fluctuating body temperatures. Torpor sets in if outside temperature sinks too low: body temperature will drop below the level allowing normal bodily functions. When outside temperatures go up, the animal's body temperature also rises, and it becomes more active.

Like many a generalization, this one suffers badly in the actual realm of experiencing nature. Many large lizards, for example, are sufficiently big that they retain heat efficiently by the sheer dint of their volume: it is smaller animals, with their high ratio of surface area (skin) to volume (body bulk) that tend to lose heat rapidly. Larger animals have far less of a problem.

And, of course, many dinosaurs were large, often huge. They and their close kin, such as the flying reptiles (pterosaurs), were *the* ecological mainstays of the terrestrial animal world throughout most of the Mesozoic. Even supposing that the ecological systems they created were not comparable in every detail to the bird/mammal systems we see around us today, nonetheless we can tell enough from their bones and teeth alone to realize that they occupied a rather wide variety of niches. And the carnivores, especially, strike all of us as capable of a great deal of activity—and not just the big tyrannosaurs and their close kin. The smaller ones are, if anything, even more lithe, graceful, and *active*-looking than their more famous, larger relatives. So the body size argument, while helpful to our visualizing how dinosaurs were so active, cannot be the whole story.

A growing number of paleontologists are realizing that the simple dichotomy between "warm-blooded" and "cold-blooded" is simply inaccurate. Why should we assume that all scale-bearing amniotes of the past—creatures we loosely call "reptiles" because they lack the advanced features of either birds or mammals—should be more similar to creatures like crocodiles and lizards than they are to, say, birds and mammals?

And this is where the argument, to my mind, really gets interesting. As we have seen in prior chapters, there is good reason to suppose that mammal-like reptiles were ancestral to true mammals, and rather mammal-like in much of their biology. (As a student, I was even shown a specimen of two mammal-like reptiles preserved together, one larger than the other, in such intimate contact as to suggest mother and child—a suggestion of parental care more typical of "mammals" than "reptiles"—and, in this instance, probably more of a fantasy than a frozen moment of motherly love.) Why could we not suppose, then, that the internal biology—the anatomy and physiology of the soft internal tissues—of various different "reptilian" groups would also vary? In particular, we might look for the nearest diapsid relatives of birds, and see if there is any anatomical reason to think that they may have been more bird-like than lizard-like in terms of thermoregulation.

So we need to ask: who among all these diapsid "reptiles," is the closest relative of the birds? And the usual answer these days, at least, is the *dinosaurs*. They simply share more features with birds—special, anatomical designs found nowhere else in the "diapsid reptiles." But this can get a bit confusing: some paleontologists, in a puckish mood, are apt to quip to a reporter that "dinosaurs are still alive today"—and then point to a sparrow chirping away in a tree.

Birds arose no later than the Upper Jurassic—the age of the oldest specimens of a true bird, a creature known as *Archaeopteryx,* which I will discuss in a moment. Dinosaurs lasted well beyond that time, so there is no way dinosaurs as a group evolved into birds. What did seem to happen is that birds arose from a species we would classify as a dinosaur—if we were fortunate enough to have a specimen of it. But we know even more than that, for there are many kinds of dinosaurs.

Time-honored wisdom sees two major divisions of dinosaurs—the Ornithischia (bird-hipped) and the Saurischia (lizard-hipped)—the division being based on the conformation of the bones of the pelvic girdle (which include the "hip"). When it was fashionable to speak of anatomical grades in evolution (remember the ammonites), paleontologists in general were very conscious of parallel evolution. Some began to doubt that there was a natural unit that we ought to be calling "Dinosauria." Ornithischians and saurischians were two

FIGURE 66 Mounting the leg of the sauropod dinosaur *Diplodocus*.

not particularly closely related groups of diapsid reptiles; they were archosaurians, along with birds, thecodonts (Triassic dinosaur forerunners), and crocodilians, but not themselves forming a true, natural group that we can call "dinosaurs." *That* term, it was felt twenty years ago, was strictly for amateurs: the resemblances between the two kinds of dinosaurs were strictly superficial: they were both huge archosaurs, independently evolved.

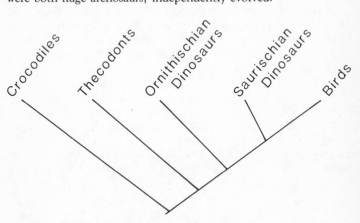

FIGURE 67   Dinosaurian evolutionary relationships.

But then paleontologists came to appreciate some special similarities that showed that ornithischians and saurischians were indeed closely related—and the old order was restored: Dinosauria came back into vogue as an official term. But it was realized soon thereafter that everything shared by the two groups of dinosaurs were also true of birds! So, if there is any utility to the term "Dinosauria," it must embrace "dinosaurs" *and* birds. Birds then "become" dinosaurs.

One irony of all this is that birds are actually more like the "lizard-hipped" than the "bird-hipped" dinosaurs. Another is that Class Aves—birds, formerly on a par with Class Mammalia and Class Reptilia—becomes a subdivision of Dinosauria, itself a division of Archosauria, which is a part of Diapsida, itself a division of Amniota.

But if the notion that birds are "dinosaurs living today" does not suggest that *Tyrannosaurus* gave rise to some early species of birds, it may very well suggest that dinosaurs possessed an internal anatomy and physiology rather more like birds than like lizards and snakes. It is hard, as we have seen, to gather evidence on soft internal anatomy from the rock record; but we do have *some* hints: it was recently shown for example, that some dinosaurs had tubes running through their bones, canals similar to the "Haversian systems" typical of mammalian and bird bones, and indicative of a degree of internal physiological control of body temperature. Sometimes "paleophysiological" inferences can be made based on detailed anatomical investigations. But analysis of relationships is often just as powerful a source of inference: once we realize the extremely close evolutionary affinities between birds and true dinosaurs, the barriers are removed from seeing how such "lowly reptiles" could possibly have behaved in the active manner so typical of birds. Taken all together, the evidence certainly suggests that the dinosaurs were warm-blooded.

We'll get back to dinosaurs when we look at the events that closed their world down. Though they dominated the Earth, they tend to crowd out of our minds all the other sorts of land and sea creatures that were also out there, including birds, early mammals, crocodiles, squamates (true lizards and snakes), plus the various groups of marine reptiles, fish and invertebrates. We need to flesh out the picture a bit just to see what was around, how it was all integrated into ecosystems, and which creatures ran into that terminal Cretaceous catastrophe.

Creatures of the Air—Analogies and Homologies

Of all the pieces of evidence that suggest that life has had a long evolutionary history, one of the most graphically compelling is the simple sameness in basic design that underlies the structural body plan of many different organisms. My own nine-year-old was struck by the repetitive design of vertebrate skeletons: noting the essential sameness between our bodies and that of the chicken he was devouring, he said, "Look at this! Of *course* life has evolved!"

Comparative anatomists of the nineteenth century were well aware of these common plans running through large segments of the animal and plant kingdoms. But even in pre-Darwinian days, when such patterns were explained as the natural consequence of a Creator sticking to the same basic blueprint for large divisions of life, there was concern to distinguish between what we might call "real" and "false" similarity, between "homology" and "analogy." And no anatomical system is better suited to demonstrating the differences between the two concepts than the various structures— "wings"—developed by vertebrates and arthropods to invade the air. invade the air.

No one looking at the translucent, boneless wings of a fly would ever suppose they were fashioned in the same way, from the same primordial source, as were the wings of a bird. Even in those few instances in which the wings are of about the same size—large butterflies and hummingbirds, say—the two are utterly different in every respect save their general placement on the body and their overall manner of use. Both are beaten up and down, providing both lift and thrust to propel the organism through the air. In that sense only, both sorts of structures are "wings." We see at once that wings must have had a separate, parallel development in arthropods and vertebrates. Thus wings in the two groups are analogous, superficially similar structures performing basically the same function, yet with totally independent evolutionary histories.

But the situation is a bit more complicated when we compare the wings of birds, bats, and the extinct Mesozoic pterosaurs (flying reptiles). Pterosaurs arose in the Jurassic and gigantic species are found as late as the Upper Cretaceous. Indeed, specimens recently recovered from Cretaceous rocks in Texas may represent the largest flying creature ever to have lived—with a wing span in the region of fifty feet.

Birds likewise appeared in the Jurassic. *Archaeopteryx*, known now from five specimens from the Solnhofen limestone in southern Germany, is among the most famous of all fossilized creatures. Possessing true feathers, *Archaeopteryx* nonetheless retains a number of primitive reptilian features (such as teeth and long tail) that mark it as a transitional form, an early version of true birds. Such combinations of primitive and advanced traits, the normal expectation of the evolutionary process, is nonetheless all too rarely encountered in the fossil record. Paleontologists are fond of *Archaeopteryx*, as it really is a legitimate instance of "mosaic" evolution and a perfect example of anatomical transition between birds and less advanced archosaurian diapsid "reptiles"—and thus good ammunition to use in combating creationist claims that there are "no transitional forms" known from the fossil record.

Bats—skipping ahead to the mammals of the Tertiary for just a moment—appear in pretty much their full-blown form in the Lower Eocene, some 50 million years ago. For bats, there is as yet no fossil record of intermediate forms linking the fully aerial creatures with their ground- or tree-dwelling ancestors—grist, perhaps, for the creationist mill, but evidence to a paleontologist that such major evolutionary transitions often occur fairly rapidly (taking, say, but a few million years at most) and in areas not conducive to fossilization. The earliest bats are found in lake sediments; it was because they could fly that some unfortunate bats met death over a lake, and managed to have their remains incorporated into the sediments accumulating below the surface.

It is clear that just as arthropods and vertebrates separately invaded the air, the vertebrates themselves accomplished the task three separate times (not counting all the gliding lizards and mammals that have approached, but not completely attained, fully powered flight). So the wings of birds, bats, and pterosaurs are analogous, not homologous—or are they?

The answer is simple, but it tells us a lot about how nature is organized. Yes, three separate times in vertebrate history the front limbs were modified for flight. Wings are analogous adaptations in the three groups. But if we remove the outer layer of skin, with its hairy coating in bats, its feathers in birds (and perhaps its hair-like structures in pterosaurs—the evidence is not quite clear), we find a haunting similarity in the general structure of those front limbs, those things we call "wings" in the three groups. Each has the same single upper arm bone, the "humerus," also present in our own arms. In each, the humerus is succeeded by two bones (the radius and the ulna), then a wrist, and, finally, "fingers."

And the basic, repeated plan of the upper arm bones in all tetrapods (and even lobe-finned fish, as we have already seen) most certainly *is* homologous. How can these wings be both homologous *and* analogous when we have already said that homology and analogy is an either/or proposition? The answer lies in those nested patterns of resemblance automatically created by the evolutionary process: the wings are homologous *as vertebrate forelimbs*. As *wings*, those forelimbs are only analogous—and further details of their structure agree with all the other evidence indicating that those three kinds of wings had separate evolutionary histories *as wings*.

The large flat surface areas of the wings of these three groups are fashioned and supported rather differently, a sign of their separate origins. It is a perfect instance of multiple "engineering" solutions to the same basic problem of adaptation. The pterosaur wing consisted of a large membrane that stretched from the forelimb down along the

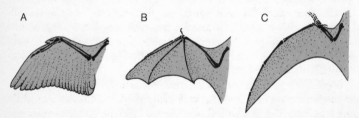

FIGURE 68   Comparison of the wings of birds (A), bats (B), and pterosaurs (C).

body. The limb had four fingers, the first three used as grasping hooks, the fourth enormously enlarged, composed of four elongated bones ("phalanges"), and running along the entire outer margin of the wing membrane.

In birds, the story is different. There are but three stubby fingers, buried in the flesh and supporting the feathers, which themselves form the flight surface of the wing. In bats we find still a third pattern of internal organization of the wing: the flight surface, as in pterosaurs, consists of a membrane of skin, but there the resemblance ends. There is a small thumb, but the other four fingers are all elongate, radiating out in a fan-like manner to support the membrane.

Thus similar structures—in this case, the forelimbs of tetrapod vertebrates—are often modified in similar, but rarely identical, ways to reach similar adaptive designs. The phenomenon is *parallel evolution*. In cases such as the history of vertebrate wings, it involves the analogous modification of more generally homologous structures. But parallelism is an amazingly common theme in Mesozoic life, and not all cases of analogy deal with homologous structures.

More Parallelisms: Mesozoic Aquatic Life

Fossils turn up in the darnedest places. Our minds almost automatically turn to distant, perhaps romantically dangerous settings when we imagine fossil hunters at work, though I have already said enough in this book to indicate that a farm in central Indiana is as likely a spot as the Flaming Cliffs of Shabarakh Usu in Outer Mongolia. It depends very much on what you are looking for.

But we do tend to think of open plains or lofty mountainous regions as prime targets for exploration; cities are where museums are—the final destination for all the hard work done in the field. But Cincinnati, Buffalo, and any number of other cities throughout the world are located smack on top of prime fossiliferous rocks, and there really is nothing intrinsic in nature, or where people have chosen to locate their urban developments, to preclude the occurrence of fossils. Even our largest American city, the Big Apple, has some spectacular fossils in its immediate neighborhood. (New York's twin city, Manhattan, Kansas—the "Little Apple"—sits on very fossiliferous marine Pennsylvanian rocks.)

New York is built on the worn, metamorphosed roots of a mountain system last thrust up in the Paleozoic. No fossils there. But right across the Hudson River, above and below the igneous Palisades sill, are Upper Triassic sediments that have yielded many specimens of fish, a peculiar gliding reptile—and one notable specimen encountered by three young geology students out exploring the base of the Palisades cliffs near the present site of the George Washington Bridge one day in 1910. There, its wet white bones glistening in the sunshine, lay what has since always been known as the "Fort Lee phytosaur."

FIGURE 69 The scrambled bones of the Fort Lee phytosaur.

Now, phytosaurs are fairly abundant and diverse in Upper Triassic rocks of Europe and North America. The main significance of the Fort Lee specimen is its unexpected occurrence: bones are rare in the New Jersey red beds, so it is a welcome surprise to have anything at all from there. Better specimens, though, there are aplenty. And when we look at complete skeletons of phytosaurs, we get another jolt: just as pterosaurs and birds sprang independently from closely related stocks of archosaurian reptiles in the Jurassic, phytosaurs and crocodiles arose from very similar Triassic thecodonts. Phytosaurs, in fact, are crocodile lookalikes. Actually, it is the other way around, for the phytosaurs became extinct at the end of the Triassic, and were replaced by crocodilians in the Jurassic: it is the crocodiles who are the phytosaur lookalikes.

There is no doubt that phytosaurs and crocodiles were ecological equivalents of one another. Crocodiles have changed only slightly since their advent in the Jurassic—another example of "living fossils" in which not much in the way of anatomical change has accrued since the group first appeared. And crocodiles appear to have had a similar ancestry to the phytosaurs, as well; crocodiles likewise seem to have sprung from some Triassic thecodont ancestors.

Yet it is clear that phytosaurs and crocodiles are independently evolved creatures. Just as extinction repeatedly took out the dominant group of ammonoids, and the stock sprang back by radiating out of close collateral kin that happened to make it through the extinction crisis, crocodiles sprang from some fresh source of archosaurians that likewise squeaked across the Upper Triassic boundary, making it while the phytosaurs became extinct. We know this—that crocodiles are not direct phytosaurian descendants—because once again, when we look hard enough, we see some consistent anatomical differences, nuances in design that show the two to have attained a body shape of crocodilian outlines independently. It is a case exactly like that of the wings of birds, bats, and pterosaurs.

Though their skulls are similar—long snouted and rather flat, with the jaws equipped with a long battery of spiky, fish-catching teeth—what at first glance seems a trivial difference is actually all the difference in the world, for phytosaurs and crocodiles came up with two different ways of solving the problem of *breathing*. Both lay low in the water. I have watched African crocodiles lying motionless for hours in midstream under low-hanging vegetation, their bodies pointed upstream, their beady eyes alert for any hapless mammals that may have fallen in, to be swept along to their waiting maws. Little but the eyes are often visible—except for their nostrils. All amniote tetrapods that have returned to the water, including the various "reptiles" as well as mammals, have retained the air-breathing lungs of their terrestrial ancestors. Crocodiles are no exception: perched out toward the tip of their snout are two external nares—nostrils. Like mammals, crocodiles have developed a second

FIGURE 70   Some unscrambled phytosaur bones: a mounted skeleton of the Triassic phytosaur *Rutiodon*.

FIGURE 71 An alligator from the Oligocene: crocodilians crossed the Cretaceous-Tertiary boundary, while dinosaurs and many other creatures succumbed to extinction.

roof to their palates (the top of the mouth). This second bony palate provides a passageway connecting the nostrils with the rear of the mouth, enabling the crocodiles to satisfy the necessities of breathing while lying semi-submerged.

Phytosaurs breathed too, but how they did so reveals their separate pedigree. *Their* nostrils were high up on the forehead on a raised protuberance. No secondary bony palate for them, they had a snorkel system leading straight down to the back of the mouth. Once again, two solutions to the same problem, even though the two groups arose from very closely related ancestors.

And, in general, when we look to marine life and see how those archosaurian and other reptiles managed to reinvade the marine environment, we repeatedly see the same sort of theme: adoption of a major habitat medium—land, air, water—time after time produces rather similar organisms. Earlier in the book I mentioned the classic case of convergent evolution—where sharks, porpoises (which are mammals) and the extinct Mesozoic ichthyosaurs all independently, from obviously different stocks and at different times in life's history,

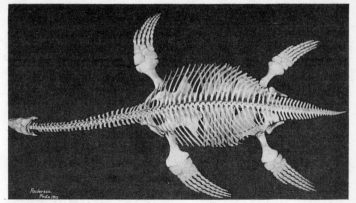

FIGURE 72 A mounted plesiosaur.

have developed a fusiform shape suitable for active predation in the sea.

But there were still other unsuccessful reptilian forays into the sea in Mesozoic times. And here we finally see the flip side to that dominant theme of repetitive invention of adaptive form. Long-necked and short-necked plesiosaurs, for example, are rather unlike any other creatures to have appeared on Earth. Prototypes of the Loch Ness monster, these creatures, equipped with paddles for swimming, were evidently every bit as effective predators as their porpoise-like compadres, the ichthyosaurs. (I have never fully understood, incidentally, why my scientific colleagues are so unrepentantly cynical, even hostile, to the very idea of the Loch Ness monster. It would be wonderful if there really were plesiosaurs still with us. The Great Glen Fault, which traverses Scotland and forms the catch basin for that incredibly deep lake, is an ancient feature that was certainly active back in the Mesozoic. I, too, am skeptical

FIGURE 73 Ichthyosaurs, porpoise and shark-like in their overall body proportions, were yet another group of aquatic Mesozoic reptiles.

that Nessie even lives, much less would prove to be a plesiosaur were we to ever get close enough to find out for sure. But wouldn't it be great if Nessie really *were* a Mesozoic refugee?)

And then there are mosasaurs, and *their* affinities lie close to varanid lizards, which still survive today in (among others) the Komodo dragons, largest and fiercest of all living lizards and even more frighteningly capable of violent destruction than a crocodile. Mosasaurs were enormous lizards (up to twenty feet in length) inhabiting the Upper Cretaceous seas. They are especially common in the chalk beds of the Niobrara Formation of western Kansas, beds which have also yielded numbers of specimens of large pterosaurs such as *Pteranodon*, as well as many giant Cretaceous bivalves (*Inoceramus*) and numerous species of teleost fish. Indeed, some of our most enduring impressions of the fabric and pulse of shallow-water marine life toward the end of the Cretaceous are in murals painted by Charles R. Knight at the American Museum of Natural History around the turn of the century, depicting pterosaurs perched on cliffs, while the waters roiled with the violent activities of fish-eating mosasaurs—an image that probably is not far off the mark.

Life Modernizes

If the fascination we all have for dinosaurs tends to blind us to the other sorts of vertebrates on land, in the sea, and in the air, preoccupation with the Mesozoic as the Age of Reptiles makes us forget about the "amphibians," the various sorts of "fish"—and the mammals. But if we spend all our time looking at *vertebrates* we will be missing most of the actual action of life. There are all those other creatures to consider: plants on land, and all those "invertebrates" on land and especially in the sea. For it is as true in the Mesozoic as it was in the Paleozoic that the bulk of the fossil record is of marine hard-shelled invertebrates: corals, bryozoans, sponges, arthropods, echinoderms—and, especially by Mesozoic times, mollusks. And if we are struck with the apparently archaic nature of the reptilian-dominated patterns of life on land, it is nonetheless true that everything else was undergoing a "modernization phase" that, with the addition of a varied array of mammals, would result in the familiar complexion of life of the last 75 million years—the Cenozoic ("Recent Life"), which most certainly includes our own era.

If ammonoids are the most conspicuous of Mesozoic mollusks, nonetheless both the bivalves and gastropods (snails) truly come into their own as major components of seafloor life in Mesozoic times. Indeed, by late Jurassic, or certainly Cretaceous times, we have for all intents and purposes the earliest occurrence of a modern-looking marine molluscan fauna. Shells preserved at many an Upper Cretaceous fossil locality look essentially modern to all but the most expert of paleontologists. And indeed, the shells are often so beautifully preserved, weathering out of poorly consolidated sands, that the unwary may well think they are dealing with the remains of last July's clambake—rather than with fossils 90 million years old.

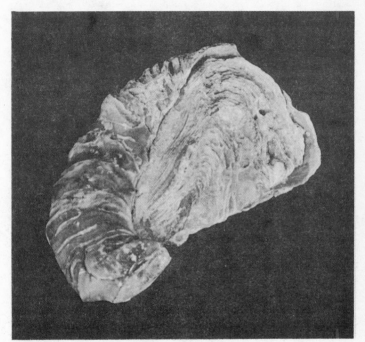

FIGURE 74   A coiled Cretaceous oyster, *Exogyra*.

There is a distinct difference in overall "feel" between the lower and upper portions of the entire Mesozoic. Early on, in the Triassic, we mainly see Paleozoic holdovers, plus extensive radiations of new groups destined to control the ecological theater at least until the end of the Triassic, and in some instances for much longer. But in group after group, families still with us today begin appearing between the Middle Jurassic and the Lower Cretaceous in a quiet but growing counterpoint to the dominance of distinctly and un-mistakably Mesozoic life forms. Thus, modern groups of cheilostome bryozoans—a diverse group of small lophophorates—begin to come onto the scene in Lower Cretaceous times. And while one can go to Jurassic localities in Europe and still collect a great number of brachiopods, by Upper Jurassic times articulate brachiopod ordinal diversity (meaning simply the number of different orders within the articulate brachiopod phylum) is down to what we see today: two orders, the terebratulids and the rhynchonellids. Today these brachiopods are restricted to the refugia of the deep sea or the crevices of rocky shores and coral reefs; in the Jurassic they were still fairly abundant in more normal, open-bottom environments; but by the mid-Mesozoic, the handwriting was already on the wall for them.

The same push toward modernization also affected bony fishes in the Mesozoic. Osteichthyan history is similar in many ways to that of the ammonoids: there seem to be definite, structural grades to the evolution of the bony fishes, as was much emphasized by paleontological students of evolution a decade ago. Thus we have the chondrosteans of the Paleozoic, the "subholosteans" of the Triassic, the holosteans of the Jurassic, and the teleosteans of Upper Jurassic through modern times. It is the teleosts that fill the seas today; they literally exploded, expanding their diversity at a reular and rapid pace, throughout the entire Cretaceous Period. Yet we have a few living remnants of the more primitive bony fish: gars and bowfins (*Amia*) are living holosteans, while sturgeons are survivors of the even more primitive chondrosteans.

FIGURE 75 *Portheus*, a gigantic Cretaceous herring from the Niobrara Chalk deposits of Kansas.

Nor were only clams, snails, bryozoans, corals, and bony fishes—all of which are animal groups—busily expanding as the Upper Mesozoic wore on. Plants were expanding too—and specifically angiosperms, the true flowering plants. Descendants of the old tree ferns and other primitive plants of the Paleozoic forests, whose remains contributed so much to the formation of the great Carboniferous coal forests, hung on in slender numbers in the Mesozoic, which was otherwise dominated botanically by the cycadeoids (relatives of the palm-like cycads that first appeared in Upper Carboniferous times), the gingkos (still with us as a living fossil) and the conifers—the gymnosperms, or "evergreens" also still very much with us today. But it is the flowering plants that now dominate the Earth botanically, and their expansion, like that of so many animal groups, began in the Cretaceous. They have been increasing in numbers ever since.

Patterns of Life's Diversity

So here we are, as the Cretaceous is drawing to a close, seeing how group after group is *expanding*—just before the Cretaceous and this chapter come to a close, and we prepare to discuss the devastating extinction event that was the termination of the entire Mesozoic. Indeed, it is in part because life within some groups, at least, seems to have been on something of an expansion kick that the extinction of so many species at the close of the Cretaceous seems so monumental.

But we must take a closer look at what we mean by diversity here. Are there really progressively more different kinds of creatures on Earth now than there were, say, in the Middle Jurassic? Has life really gotten more diverse across the board, or have some groups fallen off in proportion to the rise of others?

By diversity, most of us mean "number of species" alive at any one time. Clearly the subject is full of difficulties: the fossil record is incomplete. There is no way we can assess the true number of species alive at any moment, or even in the coarser units of geological time that we can discern given the methods available. Soft-bodied creatures are rarely fossilized at all. In fact, for all the vagaries of the fossil record, we are hardly better off when dealing with the Earth's modern biota: no one knows for sure how many species are alive today. Estimates vary from about 3 to 30 million!

But we are not helpless. We can concentrate on hard-shelled groups of animals that seem to have left a pretty dense fossil record. We can, for example, assess the number of clam species now living and come, we think, fairly close to the actual number. We can then turn to the Pleistocene (the Ice Age—the past 1.6 million years) and the Pliocene immediately preceding it, and see how many of these species we can find in the upper part of the fossil record. This will show us that for some groups, our fossil representation is pretty good.

And we can count genera, or families, as a substitute for counting species—on the grounds that while we cannot hope to find all of the species that ever lived, counting genera or families (which usually contain many species) makes the chances much greater that we will find at least one species belonging to each of these higher sorts of units. Paleontologist David Raup of the University of Chicago has been especially active in tackling the vexatious problems in the counting up of life's present and past diversity.

There are two extremes of opinion on the diversity of life over the past 575 million years. Some paleontologists think that the pattern that seems fairly common from the Upper Mesozoic through the Cenozoic—steadily increasing diversity—has held true throughout life's entire history. The thought here is that life continues to become more complex, slowly but steadily filling up all the ecological opportunities that can possibly be imagined and realized. Once something new evolves, it changes the nature of the game for everything else—and while extinction may be the result for some, there might even be further opportunities created for still other organisms. In any case, according to this scenario, life is not limited to a finite number of ecological niches.

Others have said that the Cretaceous-Recent increase in diversity merely reflects the burgeoning success stories of isolated groups—and that life in general, if it could all be added up, would show no net increase in diversity over that period. Instead, many paleontologists have felt that life's diversity has been in a state of near equilibrium ever since that first explosion in the Cambrian. In this view, it doesn't take life long to fill up an ecological vacuum—opportunities created by extinction or by the invasion of totally new environmental situations, such as land, or the air, or the reinvasion of the sea. Life does not dilly-dally around: it exploits all opportunities to the hilt and does so in a few millions of years. There is no reason to expect life to take hundreds of millions of years, slowly filling up the available niches.

The latter point of view very much appeals to me, as earlier discussions in this book make fairly clear. But what about the facts of the matter? Well, I am no fan of compromise, but it is interesting that several of the more active students of life's diversity now agree that life seems to have been in equilibrium pretty much since the end of the Cambrian diversification (with strong fluctuations caused by the major extinction events). But since the Jurassic, they also agree, life has definitely been on pretty much of an expansionist phase—and the terminal Cretaceous event seems hardly to have slowed the pace of diversification.

Well, perhaps. The fossil record gets better and better, the rocks more fully represented, the closer we get to the Recent. Still, it is hard to deny the numbers. And below the facts and the problems in trusting the data lies a much deeper question: is there really a finite number of ecological niches on Earth? We really do not fully understand the patterns of diversity on Earth at this moment: for example, the tropics house vastly more different kinds of animals and plants, on both land and sea, than can be found in the higher-latitude temperate and arctic regions. Why? It seems that there are nearly as many ideas to explain the pattern as there are biologists. One thing everyone agrees on: there is more energy available in the tropics than in the higher latitudes, which receive more diffuse sunlight, therefore less energy, during half of the year.

But is evolution faster in the tropics, or is extinction simply slower? If the former, the tropics are a laboratory of evolution; if the latter, they are more like a museum, simply harboring relics of ancient times. Africa today has by far the richest land mammal fauna, but in large measure the disparity between Africa and, say, North America reflects the fact that far more of the Pleistocene mammals of Africa managed to escape the sweeping extinctions that so recently annihilated so much of North America's mammalian fauna.

Understanding the vicissitudes of life's diversity—the old game of matching up ideas with actual observation of events in the history of life—has proven to be no easy matter. We turn now to an event, the terminal Cretaceous extinction, that presents us with a few pretty clear facts, but also with some ambiguous data. And the problems of matching ideas on the causes of mass extinctions with what we can actually observe are either fascinating—or bad enough to make a strong paleontologist weep.

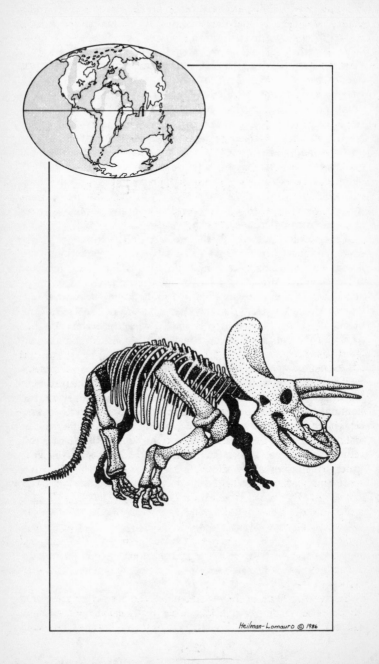

7

Extinctions: Resetting the Evolutionary Clock

Many years ago, long before I became a professional paleontologist, I read of a cowboy riding the range in some godforsaken arroyo out west. When he finally got back to the ranch, he told of huge bison horns jutting from the slopes where he had been riding along, and of eye sockets so massive he could toss his ten-gallon hat right through them.

I have no idea whether the story is true. But its moral is accurate enough: those were no bison that cowboy saw from the saddle. They were *Triceratops* skulls—belonging to those great, tank-like (or, more simply, rhinoceros-like) ceratopsian dinosaurs. The holes were not eye sockets but the large openings on the side of the head—the same holes that mark all dinosaurs as diapsid reptiles. But what is really striking about these fossils is their *commonness*: geologists really do traipse around the uppermost Cretaceous rocks in Montana and map their rocks by the frequent scraps of *Triceratops* still littering the surface, long after that cowboy started punching cows on the Great Ranch in the Sky.

Stories like these, of course, add fuel to the fire of that greatest of all mysteries of dinosaurian lore: what really *did* happen to those ancient reptiles, anyway? Their demise at the end of the Cretaceous has stirred our collective curiosity for well over a century, and has goaded our imaginations to some pretty remarkable heights—or depths. But never has scientific interest in mass extinctions in general, or in dinosaurs in particular, been greater than it has been in recent years. When in 1984 *Time* magazine sported *Tyrannosaurus* on its front cover and asked "Did Comets Kill the Dinosaurs?", we knew that extinction had hit the big time, and that dinosaurs were right in the thick of things. That's why I have left a general discussion of mass extinctions to the transition between the Mesozoic and Cenozoic: the extinctions that mark the end of the Cretaceous were not the largest (the Permo-Triassic extinction was even greater), but they are the most famous, perhaps the best studied—and the focus for most of the latest batch of theorizing. Thus a consideration of this most famous of all extinction events has much to tell us about extinction events in general.

Extinctions: Some Basics

Before we zero in on the end of the Cretaceous, we should recall some of the basic threads of our narrative of life's history. We have seen throughout a duality in the basic features of life itself: economic matters, in which creatures are concerned with obtaining energy simply to grow and to live, and genealogical matters, in which genetic information is stored and transmitted in reproduction. Evolutionary change most often entails a modification of adaptations for living: the behavioral and anatomical properties organisms use simply to make a living. The changes are recorded by natural selection in the reservoir of genetic information. What works in the economic sector is written in the DNA of life's ledger book, to be retained for the next generation and for millennia to come unless the evolutionary clock is somehow reset.

And so we naturally ask, how is that clock reset? One idea derived from Darwin and duly incorporated in the canon of modern evolutionary theory is the notion that old solutions to economic problems eventually must give way to newer, better engineering designs. Natural selection is seen as a force constantly winnowing and sifting, searching for improvements in basic design. Darwin wrote in an age when progress and improvement were commonly considered inevitable outcomes of competition—and progress is still very much built into our expectations of evolutionary history. As a result, the usual explanation for the demise of the dinosaurs is that at long last something better—*mammals*—came along and took their place. And we have already seen that such a notion is false on its face—that mammals were there throughout at least the latter two-thirds of the

Mesozoic. Hardly superior creatures, they were the "rats" of the Mesozoic, as Alfred S. Romer put it. The pattern looks far more like a simple matter of opportunity: only after the dinosaurs had been dislodged from the great assortment of ecological niches they filled could the mammals get going and themselves radiate into a series of analogous ecological vacancies.

And we have also seen enough to know that extinction is not primarily a genealogical affair: it is not a matter of mammals replacing dinosaurs, or clams becoming numerically predominant over brachiopods as the Triassic dawned after the crisis that terminated the Paleozoic. It is much more an ecological affair—in which a variety of only vaguely related groups are all knocked off at about the same time. Thus many brachiopod groups, including the oddly shaped, specialized reef-dwelling brachiopods, disappeared along with trilobites, rugose corals, many snails, nearly all ammonoids and so on at the close of the Permian. And on land the Permian story, particularly in the Northern Hemisphere, was much the same. What is arresting, then, is that on land and in the sea, air and freshwater, with plants and animals of diverse phyla, plus minute single-celled organisms (creatures that leave their mark as microfossils)—and presumably even bacteria—we get these coordinated, and sometimes incomprehensibly massive, dyings. They are nothing short of calamitous degradations of ecosystems. The very biggest involve the entire world biotal system. And one message is already clear: the issue is first and foremost *economic*. Something goes wrong with the energy-procuring side of life's activities. We measure the *results* in genealogical terms; we tabulate the demise of this group and that, the dinosaurs, or the ammonites. But we are merely using nature's own accounting system to report the score. When it comes to understanding what was going on, most of us are by now attuned to thinking of ecological crises as somehow lying at the root of the problem.

Extinction Theories: Some Older Notions

It is easy to fall into the trap of "Whig History"—in which earlier figures look silly because their ideas turned out to be "wrong," at least by present standards. But it is the very essence of science to conjecture about the world and then to go out and see if the world matches up with these expectations. Sometimes it does, only to have later generations laugh our fondest notions right out of court.

Seeing mass extinctions as ecological affairs immediately eliminates any *internal* explanations—the idea that something went wrong with a particular group. Past generations of paleontologists saw nothing intrinsically ludicrous about the notion of "racial senescence" (encountered in the previous chapter in connection with the ammonites). Today we think that the analogy between large groups (species and still larger groups) and single organisms holds only to the point of seeing them all as sorts of "individuals;" there is no reason to push the analogy too far and claim that species undergo an aging process even remotely similar to the aging of organisms. And in any case, such theories are "special case" notions—totally ignoring the fact that the extinctions in question cut across genealogical lines. No sense asking what was "wrong" with the dinosaurs when virtually every other family of organisms was also undergoing major upheaval.

But some attempts to understand extinction in environmental terms suffer from a lack of catholicity: they just don't cover enough different kinds of animals and plants to suffice as truly general modes of explanation. My favorite recent example is the suggestion that dinosaurs died off from selenium poisoning. Ingestion of selenium in small doses, it is said, can be good for you; it also may make you smell like a goat. In larger doses, it kills you. The Mowry Shale, a Cretaceous marine unit of the western United States, has been found to contain selenium in quantities far higher than levels in most other sedimentary rocks. The elevated concentrations of selenium in the Mowry are usually ascribed to heightened volcanic activity in the region. Perhaps, according to the theory, the selenium got into the food chain via plants, as it has been known to do in modern times. Perhaps herbivorous dinosaurs were poisoned.

Perhaps. But such volcanism was hardly worldwide; the Mowry did not come at the end of the Cretaceous (far from it: the Mowry occurs about halfway through the Cretaceous Period). And even though looking at the base of the food chain is a promising start for a truly ecological theory of extinction, the hypothesis seems to be geared to terrestrial organisms, specifically dinosaurs, and is thus too narrow to give us that truly general explanation that such events seem to require.

Then, too, when DDT was found to weaken the calcification of bird eggs (precipitating the rapid decline of Osprey populations along the Atlantic coast of the United States—a situation that has recently and fortunately been reversed), some paleontologists speculated that dinosaur eggs had been weakened by some naturally occurring toxic substance. In the late 1960s and early 1970s, when environmental concerns were perhaps at their peak (and when "ecology" largely signified pollution), it was natural to see past extinctions as pointing to the possibility that mankind might be engineering yet another disastrous episode. But, again, focusing on one group—especially the most conspicuous group of all, the dinosaurs—is hardly the route to take toward a general theory of extinction. Nor is disease a sound concept on which to posit extinction: the rampant path of some disease-causing microorganism could conceivably take out an entire species. And that species might be integrated into more than one local ecosystem, with the result that its disappearance might prove inconvenient, if not completely fatal, to some other species. But once again, we are hardly contemplating the stuff of large-scale, cross-genealogical disasters.

Toward a Valid Theory of Mass Extinction

Though extinctions provide the basis for both the major and minor subdivisions of the geologic time scale, hardly anyone paid them much heed for years. Virtually alone as a voice calling out in paleontology, Norman D. Newell (for many years at the American Museum of Natural History and mentor of, among many others, Stephen Jay Gould and myself) sought to alert his colleagues to the importance of extinctions. Newell spent long months patiently accumulating data documenting the faunal turnovers at the end of the Permian and Cretaceous Periods. His efforts at data gathering and analysis—mostly at the level of entire families—directly foreshadowed today's mammoth computerized data banks, refined to a high art by John J. Sepkoski of the University of Chicago.

Like all good scientists, Newell was collecting his data for a purpose. He wanted to document the reality of extinctions. And he wanted to demonstrate their severity, to show with actual numbers what he felt must be true: that the biggest of these "crises in the history of life" (as he called them) very nearly wiped out all life. But at the very heart of his labors lay his ultimate goal: to reach an understanding of the actual *causes* at work, the real reason why so many creatures dropped out, all at pretty much the same time.

Newell knew he had to come up with some general environmental theory, something that could integrate extinctions both on land and in the sea, something that would be largely indiscriminate, not selectively picking on just a few kinds of organisms. That something would probably turn out to be an aspect of the physical environment.

An invertebrate paleontologist, Newell had turned his attention to the Permian rocks and fossils of the American Southwest (see Chapter 5) in earnest by the early 1950s. The very top of the marine units in the Delaware Basin of Texas is disappointingly barren of fossils: the so-called Ochoan beds are mostly salt. Looking elsewhere around the world, Newell thought he saw evidence for a worldwide standstill toward the end of the Permian: the seas were largely withdrawn from the continental interiors, and erosion of the continental interiors seems to have been reduced. The continents, in other words, had been eroded to fairly low, featureless plains by the end of Permian times. Looking up to the end of the Cretaceous, Newell thought he saw pretty much the same thing. The widespread chalk deposits of the Upper Cretaceous seemed to him to imply quiet conditions, with little runoff from the lands. (Chalk is composed of bits and pieces of invertebrate shell material, but consists mainly of untold numbers of tiny plates from calcareous algae known as coccolithophorids). Here again Newell thought he had reason to suspect conditions along the marine margins of the continents. Chalk will form only when erosion of the land slows to hearly a halt, and few impurities (muds, silts, and sands) are being deposited.

From his marine perspective, Newell saw that areal space—the simple amount of habitat available—shrank when seas were withdrawn from the continents. That in itself could cause crisis conditions. And reduced runoff from the continents implied a decrease in the amount of nutrients washing into the seas—nutrients (such as phosphorous and nitrates) essential for marine organisms that otherwise obtain them mainly in areas of upwelling, where water circulation patterns bring them up from the depths. (Areas of upwelling in today's oceans correspond to the world's greatest fisheries areas.)

Seemingly fitting right into Newell's scheme was the data provided by those minute coccolithophorids, which showed that tens of species had simply dropped out suddenly at the very end of the Cretaceous, to be replaced by another set of species in the Lower Paleocene. (See Figure 79.) These armored algae were a clue to what was going on at the very base of the marine food chain, and evidence that there was a general collapse of the marine energy pyramid. Paleontologist Thomas Schopf of the University of Chicago calculated the shrinkage of habitat space in the Upper Permian—the locus of Newell's initial interest in mass extinctions—and corroborated Newell's thesis that the living space, for shallow-water marine creatures, precisely the sort of organisms best documented in the fossil record had declined drastically. Since the late 1960s, scientists have realized that diversity, in terms of the number of different species of organisms living in any one area, is directly correlated with area: the greater the area the more species. Moral: reduce the area and you get fewer species. Curtail the area drastically and you are likely to get a crash.

Such theories are compelling, if only because they rely on physical factors and offer plausible reasons for the degradations of truly large-scale ecosystems. There was one major drawback to Newell's suggestions, however: the link of the marine shrinkage and food crisis to similar events occurring at the same time on land was not as satisfactory as might be wished. But in its favor, the explanation relied strictly on the sorts of events known to occur with some regularity in the Earth's history. Nothing mysterious had to be invoked. Note, too, that changes in sea level can occur at "intermediate" rates: not overnight, to be sure, but quickly enough on the order of thousands or tens of thousands of years to produce what *looks* like a relatively sudden event when we examine the fossil record.

Recently, with the advent of Plate Tectonics, we have a better idea of what might cause worldwide oscillations in sea level. When plate motion is relatively active, new crustal material is being generated over widespread areas. Mid-ocean ridges bulge out as seams along the oceanic floors, and these seams are voluminous: they displace a lot of sea water, which has to go somewhere. It spills over onto the continents. Moreover, mountain building is very largely the effect of plate motions; the collisions of plates wrinkle up the Earth's crust where two plates come togehter. And uplift leads to erosion, which leads to nutrients being cycled back down to the sea. Thus, a high degree of plate activity should result in a fair amount of evolution; conversely, it may be during times of low activity, when things grind to a halt, that events conspire to yield extinctions.

But there is another natural, earthly mechanism accounting for most of the worldwide oscillations in sea level: the great polar ice caps. Today, even though we are in an interglacial phase, the caps are larger than they have usually been over the past 600 million years. This means that the continents stand high and dry—a rare circumstance during the history of life. Ice Ages are very much implicated in the extinctions that have taken so many large mammals out of the world's ecosystems as recently as the last 10,000 years—though the hand of man is suspect as at the very least a coconspirator with Mother Nature in the demise of many of these creatures.

But it is not so much ice sheets themselves as the worldwide lowering temperatures that they imply that arrests the attention of a paleontological detective in search of the root causes of mass extinctions. One of the oldest speculations on the "death of the dinosaurs" is a drastic decrease in temperature. Dinosaurs, remember, are conventionally deemed to be "cold-blooded"—by dint of their "reptilian" status. Yet even though we have cause to think that at least some of them were functionally endothermic (see

Chapter 6), this is no guarantee that a worldwide cold snap would not play hob with their existence—or with the existence of every other species around. Just because an idea has been in existence for a while is no indication that we should think any the less of it. It had also been proposed that the Earth got too *warm* for the dinosaurs and other Cretaceous creatures, despite the lack of any evidence for this. But we now are getting some solid geochemical data supporting the idea that toward the end of the Cretaceous, the Earth really did get colder than it had been.

Indeed, the world in Upper Mesozoic times seems to have been a rather uniformly mild place. There was little to distinguish the tropics from the temperate latitudes. Given this, let us now follow paleontologist Steven M. Stanley's line of reasoning: what would happen to organisms if the temperature were to decline worldwide by an average of just a few degrees centigrade? In keeping with the principle that organisms tend to move with their habitats, we would expect higher-latitude species to move toward the Equator; we might also expect extinction to claim a number of species already clustered around the Equator, as they simply would have no place to go—no haven where the temperatures to which they were adapted continued to prevail.

Stanley reports a "natural experiment" of sorts that bears on this scenario and seemingly corroborates it. During the Pleistocene Epoch—the past 1.6 or so million years—while the Northern Hemisphere was experiencing at least four major episodes of glacial advance, the marine invertebrates on either side of the North American continent experienced rather different fates. According to Stanley, the extinction of Pleistocene mollusks was very high on the Atlantic coast, as patterns of circulation in the North Atlantic simply prevented many species from retreating southward toward warmer waters. In contrast, circulation patterns on both sides of the Pacific allowed free interchange of habitats, and organisms by and large were much more successful there in keeping track of comfortable water conditions. The result: a far smaller extinction of Pliocene and Pleistocene species in the Pacific than in the Atlantic. This makes a lot of sense, seems to be supported by real data, and (once again) invokes no processes other than those known to occur with some regularity on the Earth's surface. It is true that the ultimate causes of glaciation are as yet only dimly perceived. It is also true that glacial advances may ultimately be triggered by some extraterrestrial event. And *that* possibility gets us to the big area of active research in extinction these days: the search for non-earthly causes for mass extinctions. These are the theories that are getting extinction all the current media exposure.

Not of This Earth: Extinctions as Cosmic Doom

There really can be stimulating episodes of cross-fertilization between fields of science that normally—and quite naturally—have nothing to do with one another. We have already seen (Chapter 4) how astrophysics and paleontology were successfully linked in elucidating the rate of the earth's spin—and consequently the length of the Devonian day—as it was 400,000,000 years ago. Physics made some predictions, and paleontologist John Wells was able to test them rigorously by counting the daily growth bands on some of his well-preserved rugose corals.

Nowadays the search is on to link mass extinctions with ex-traterrestrial causes. It is a fascinating exercise in modern science. Though I harbor doubts about the results, it is certainly true that the story that emerges from all this activity really does contain the in-gredients of *bona fide* science. There is hard evidence supporting the notion of collisions between the Earth and some sort of foreign objects—be they asteroids or comets. There is even the prediction arising from paleontological data that mass extinctions are periodic (meaning cyclical, repeated at fairly constant intervals). Ex-planations for this supposed cyclicity include the effects of the rota-tion of our entire galaxy (triggering periodic collisions with ex-traterrestrial bodies). But they also include the overt prediction that there really is an as yet undetected tenth planet in the solar system ("Planet X") or, more glamorous still, a· sister star to our own sun—one that periodically dislodges comets and sends them on a collision course with Earth. Paleontologists have actually set some planetologists and astronomers to work on projects they otherwise would never have dreamed of pursuing—looking for Planet X, or for Nemesis, the putative dark and deadly companion star to our sun. All this is very exciting. The work is being touted as the breakthrough we have all been looking for in our understanding of mass extinctions. And even though there is no agreement about *which* of the various scenarios is correct, many physicists and paleontologists seem sure that the answer to what causes these environmental crises lies somewhere in this line of thinking. I am inclined to doubt it, but before I reveal why, we should look more closely at this research.

The current flurry of interest in extraterrestrial causes for ex-tinctions stems from the research of geologist Walter Alvarez, who was studying the rocks at the Cretaceous-Tertiary boundary that are especially well preserved at Gubbio, Italy. Luis Alvarez, Walter's father and a Nobel Prize-winning physicist, came into the picture when together they reported a puzzling anomaly in the data. In a thin band at the very top of the Cretaceous rocks at Gubbio, there was a concentration of the rare element iridium—some thirty times richer than the iridium content of most other rocks. Question: what could have produced that off-scale accumulation of iridium in that one-inch-thick band of red clay right at the Mesozoic-Cenozoic boundary?

Iridium is found in such concentrations deep inside the Earth. And it comes in high doses in meterorites—which are fragments of asteroids, most of which are confined to an orbital band between Mars and Jupiter. When similar iridium-rich layers turned up elsewhere, and always right at the boundary between the Cretaceous and the Paleocene (i.e., the earliest epoch in the Tertiary Period), the Alvarezes and their colleagues knew they were on to something. Discounting volcanoes as a likely source, they postulated a gigantic asteroid impact—a collision between Earth and a huge chunk of material from the asteroid belt. And of course, such an event would have ramifications. Indeed, it was not so much the impact *per se* but rather the huge dust cloud that would have resulted that next attracted their attention. Knowing that major volcanic eruptions regularly send up dust clouds that encircle the Earth in the upper atmosphere, coloring sunsets for several years, the Alvarezes postulated such a dust blanket so many more times dense than the greatest volcanic clouds that the Earth's surface would have been cut off from most of the sun's energy. Photosynthesis would have ceased, and plants and animals would have died—in short, here was an ideal trigger for that devastation, that worldwide ecological collapse. Very much the same sort of concept underlies the notion of "Nuclear Winter"—the scenario of what might happen to the Earth's atmosphere should a nuclear war occur.

There are some intriguing items that seem to back up the Alvarez scenario. Paleobotanist Andrew Knoll of Harvard University has written of pollen and spores in the fossil record at the Cretaceous-Tertiary boundary. Knoll points out that in adverse circumstances, vascular plants can be drastically cut back to the point where they cease producing pollen, but that their root systems often survive, so that growth very quickly springs back to normal. Right above the iridium anomaly, pollen drops out; only spores (from ferns) are in evidence. Soon thereafter, pollen returns—every bit as if a drastic devastation had been weathered, and vascular plants had sprung back into life—to the point where they could resume the luxury of their reproductive activities once again.

But some geologists have preferred volcanism to an extraterrestrial origin for all that iridium. Others have pointed out that the impact source might just as plausibly have been comets as asteroids. Indeed, comets are now the favored candidates because for some, at least, it is easier to envision how comets might come in regular cycles than it is to think of periodic storms of asteroids. Though we do have meteor showers occurring like clockwork every year, the sort of periodicity that some paleontologists have recently been claiming for mass extinctions is on the order of tens of millions of years. Paleontologists David Raup and Jack Sepkoski, both of the University of Chicago, have compiled the best data and posit a cycle of some 26 million years. A statistical analysis of all the data that Sepkoski has amassed on families in the fossil record really does seem to substantiate the cyclical repetition of mass extinction events—some bigger than others, of course. The correlations are not perfect, but they seem to be good enough to have inspired the search for a cause of the extinction. Mankind clung to the notion of rotational cycles in the heavens for thousands of years—and for good reason. Newtonian mechanics works reasonably well in such simple systems, well enough that it wasn't until early in the twentieth century that Einstein showed that those mechanics are incomplete and partially incorrect.

Hence the search for a death star, aptly named Nemesis, sister to our own life-nurturing Sun. Or perhaps it is the missing Planet X that periodically induces comets to collide with the Earth. Or maybe, as our Earth bobs up and down through the flattened "plane" of our own galaxy, it passes through clouds of dust that somehow set off cometary collisions. Word is that all of these various planetological and astronomical hypotheses are in severe difficulty. All appear to be on shaky ground—which of course does not make them wrong.

What we are seeing is fairly typical in science. Enthusiasm for a theory builds because some preliminary data and analysis seem to fit one—or an entire set of—working hypotheses. There are always skeptics, all along the way. Eventually the "truth" of the matter comes out in the wash (truth is seldom if ever arrived at in science: you may have the truth, but it is virtually impossible ever to be sure that you do!). This is not to say that extinctions are not caused by extraterrestrial impacts, or that mass extinctions are not periodic. Periodicity has been invoked from time to time in Earth history, and in the history of life, as long as there has been the scientific study of these subjects. It usually enjoys a brief run in the sun, to be later obscured as a decided majority rejects it for one reason or another.

But it is not because of the increasing disillusionment among astronomers with the search for Nemesis or Planet X that I harbor my own doubts about the search for the extraterrestrial causes of mass extinctions. Nor do I automatically recoil at suggestions of periodicity, though I think that were these events truly caused by comets, they would be rare enough that we would expect them to be scattered in time, giving us an average frequency more like the rate of atomic decay than a reflection of a rigidly cyclical process. No, I have other, paleontological reasons to doubt these extraterrestrial scenarios.

Years ago, when I was still in graduate school, I took a week to prepare a seminar for none other than Norman D. Newell. The subject: the mass extinction at the end of the Cretaceous. The data were a lot harder to dig out of the scientific literature back then than they would be now. There are more handy compendia to consult now than there were back in the mid-1960s. Nonetheless, dig I did, and I cannot forget what I found: (1) yes, there really was a prepossessing, cross-genealogical extinction of rather stupendous proportions. Very few of the species, genera, or even families found in Upper Cretaceous rocks were to be found in Paleocene rocks. But (2) the extinction event was well underway long before the "Golden Spike" was driven into the ground to form that Cretaceous-Tertiary boundary. The uppermost zone, in which marine conditions persist to the very end of Cretaceous times, lasted for about 1 million years. Though it is true that extinctions were very much accelerated within this last million years of the Cretaceous, the extinction was long underway millions of years before the Cretaceous had drawn to a close. Many families had been long gone before whatever it was left all that iridium lying around. Nearly all the dinosaurs were certainly gone—only *Triceratops* and a few others were around at the bitter end.

Among the ammonoids, only the Scaphitidae—the oddly coiled creatures that in part inspired the fantasies of racial senescence—seem to have still been in the throes of expanding. On a graph of diversity, it certainly looks as though the scaphites were cut off in mid-stream of their evolutionary history. But this was not so for most of the rest of their fellow earthly inhabitants.

But I readily admit that I am no expert on Cretaceous fossils. My experience as a graduate student merely biased my reactions against extraterrestrial scenarios. For the data all indicated—insofar as I was able to judge—that *had there been such a collision, it could not have been the trigger that set off the extinction.*

Paleontologists who are experts in Cretaceous fossils—Erle Kauffman of the University of Colorado at Boulder, who works on Cretaceous mollusks of the western Cretaceous inland seas, and William Clemens of the University of California at Berkeley, who is a vertebrate paleontologist specializing in Cretaceous fossils—are quick to point out that the extinction was well under way *before* any impact (if such it really was) ever occurred. I agree with S. M. Stanley: the role of an impact was to make a bad situation truly awful.

We have been witnessing, I think, an understandably enthusiastic, but still off-scale, over-zealous jump onto a bandwagon. The arena has been the plausibility of this or that astrophysical scenario—with not enough attention to the basic question: what, really, is the pattern of that monstrous extinction event? All the astrophysically inspired scenarios see the extinction in terms of minutes, hours, days, weeks, or, at most, a few tens of years: in any case, not in terms of the tens or hundreds of thousands of years (not to mention one or two million years) that actually passed while the Mesozoic biota underwent their decline and fall. Thus the "event" was a protracted decline—quick in geological terms (a million years is only .55 percent of the total 180-million year duration of the Mesozoic)—but hardly the stuff of an overnight event.

Ecological degradation, especially at the scale of the entire world, land and sea, microbe to vertebrate, is not likely to yield to anything nearly so simple as a smack on the jaw. We really are dealing with the collapse of ecosystems here. Ecosystems are complex. Something must have triggered a degradation of these complex systems in order to cause these mass extinctions—something, it could be agreed, like an impact—though the data show such events come well after the extinction "event" is long underway. More probably, something like a "sudden" (thousands of years?) drop in worldwide temperature was what set off that subtle but cascading chain of events. The events that went furthest get logged as the biggest of all extinctions. Perhaps they really were booted along by extraterrestrial impacts. Perhaps not.

But let us not forget that mass extinctions are *real*. Life really did take it on the chin *severely* many times—especially at the end of the Cambrian, in the Upper Devonian, at the end of the Permian, the end of the Triassic, and at the end of the Cretaceous—and less severely so at many other times. Geologically, I think it looks instantaneous, but realistically, it really isn't so. And these events, whatever their cause, whatever their pattern of occurrence, whatever their true time frame, really did reset that old evolutionary clock. They really did create those opportunities to rejuvenate, for the truly new to spring forth. Which brings us to the mammals—ourselves, among many others—who inherited the Earth after that big, but not so bad, terminal extinction that brought the Mesozoic to a close.

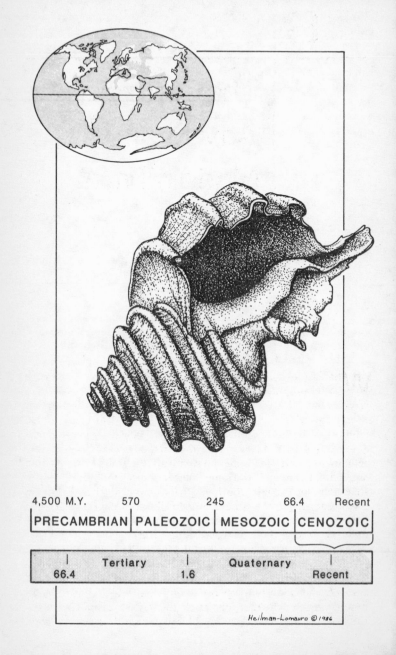

4,500 M.Y.	570	245	66.4	Recent
PRECAMBRIAN	PALEOZOIC	MESOZOIC	CENOZOIC	

	Tertiary		Quaternary	
66.4		1.6		Recent

Heilman-Lomauro © 1986

8

The Cenozoic: Advent of Modern Times

Whatever the factors were that carried off such a huge percentage of the Earth's species toward the close of the Cretaceous, the stage was set for opportunistic takeover as ecosystems once more settled down to business as usual. Some groups—crocodiles and turtles, for example—made it right through. And many of those that did cross the Cretaceous-Tertiary boundary remained much the same anatomically, hardly changed and represented by roughly the same number of species as before. Horse-shoe crabs, for example, continued to look pretty much as they did in the Mesozoic, and the rather scanty fossil record suggests that there were probably no more than four or five species alive at any one time from the Triassic right up through the Recent.

But by now, we are familiar with the theme of radiations, in which groups show a burst of evolutionary activity, an increase in numbers of species together with a proliferation of anatomical shapes and sizes, often coming as a response to opportunities. And invasion of new habitats, or the repopulation of the world following major ex-tinctions, is the major source of economic, thus evolutionary, opportunities. The rise of the mammals after the Cretaceous is a great example in which this script was followed: the meek really did inherit the Earth, and even some of the details of that radiation correspond closely to patterns we have already encountered with other groups in earlier times. Nor were the mammals the only creatures to undergo such a burst: very much the same sort of thing happened in the sea, as modern life finally began to take shape. And that is what "Cenozoic" literally means: "recent life."

Fossils and the Division of Cenozoic Time

Charles Lyell (1797-1875) is generally given credit for founding modern geological practice. It was he who stressed the need to understand the geological past in terms of processes we see going on around us today, furthering the earlier work of James Hutton (1726-1797) in the late eighteenth century. Lyell, for example, argued that some curious hills in central France were really ancient, extinct volcanos; they could be understood by comparing them with known (still active) volcanos. And it was Lyell who named the divisions of the Tertiary Period (the vast bulk of the Cenozoic; the "Quarternary" consists of the Pleistocene and Holocene, which together account only for the most recent 1.6 million years).

Lyell classified Tertiary rocks according to how close their fossils resembled modern faunas. He saw a progressive similarity between fossil and Recent faunas that, naturally enough, increased the closer one got to modern times. He devised a simple percentage method of calculating that similarity: thus his "Eocene" (then the earliest division) was the "dawn of the Recent" and had, by his reckoning, only 3.5 percent of its species in common with the Recent. And so on, up through the Miocene ("Middle Recent"—with less than 18 percent), and Older (33 percent to 50 percent) and Newer (90 percent) Pliocene ("nearly Recent").

Darwin, himself originally far more a geologist than a biologist, took Lyell's new book *Principles of Geology* (published in three separate volumes in the early 1830s) along with him on the *Beagle*. Lyell's arguments helped convince Darwin that geologic time was vast, certainly long enough for evolution to occur. Long enough, that is, if evolution was really the slow, gradual affair that Darwin conceived it to be. And it is ironic that long after Darwin's *Origin of Species* (1859) had been published, Lyell was still steadfastly opposing the very idea of evolution. The irony comes in because it was Lyell's central idea that geological processes were inherently *gradual* that intrigued Darwin so much. Lyell had made the notion of a gradual change through time popular; Darwin adopted it as a cornerstone of his evolutionary theory, only to suffer the disappointment of seeing Lyell deny that evolution occurred at all.

And yet Lyell had his "percentage similarity" index for comparing Tertiary fossils—hence rocks. Surely the most sensible interpretation of such data would be evolutionary change, a modification of organisms progressively as we climb the rock sequence and get ever closer to modern times. But Lyell was a "steady-stater"; the world, he thought, was always very much the same sort of place we find it to be today. That, he thought, was the very basis for our belief that we can understand the past in terms of processes we see going on around us today. He even insisted that mammals must have been present in the Paleozoic. He was understandably elated when mammals were discovered in the Jurassic Stonesfield Slate in England during his lifetime. But his denial of a "progressivism" of life through time was not so strong as to blind him to the gradual approximation to the Recent shown, in particular, by marine molluscan fossil faunas in Europe. Today, his scheme has been abandoned, though his names for Tertiary subdivisions are very much alive. It turns out that a simple percentage method won't work because faunas changed at different rates in different climates. As we saw in the last chapter, the mollusks on the Atlantic coast suffered severe extinction during the last Ice Age, while many species alive today along the Pacific coast of North America survived intact from Pliocene times.

Mammals Come into Their Own

Lyell's use of mollusks to divide up Cenozoic time might lead us to look for the same pattern in mammals as we begin to consider their spectacular radiation. But any expectation of gradual evolutionary progression, of slow development from the Paleocene up through the Pleistocene, will quicky be shattered by the actual mammal record. In a nutshell, it didn't happen that way. Mammalian evolution in the Paleocene showed a characteristic lag; it was not until the Upper Paleocene that a full-fledged fauna had developed. And then there was a rather sharp, marked extinction event: many of the "archaic" mammals abruptly disappeared. It was only in the Eocene that most of the truly modern groups of mammals began to appear.

Radiations are usually not simple affairs. They typically involve an initial burst of one or two lineages following the usual lag after a mass extinction; then there is usually another extinction that resets the clock, paving the way for collateral lineages to take over and produce a radiation of their own. That's what happened with the mammals.

When the early bone hunters hit the trail to the western plains of North America—and beyond to the basins scattered among the ranges in the Rocky Mountains—they weren't after dinosaurs alone. They soon discovered that rare mammal fossils (mostly in the form of teeth) were in those Mesozoic rocks, too. And being human, paleontologists are often most attracted by the rare material. That and the simple observation that we ourselves are mammals explains much of the fascination paleontologists have had for the mammalian fossil record.

Nor did Leidy, Cope, Marsh and all the others that came after them confine their attention to Mesozoic mammals: they really are quite rare, and fossils so few and far between could not long have occupied the attention of men with such voracious appetites. Most of "Cope's Bible," that stupendously thick compilation of E. D. Cope's scientific work published in 1883, is actually devoted to Tertiary mammals. The mammal-bearing Tertiary rocks of the American West were every bit as much of a vertebrate paleontological treasure trove as were the Mesozoic rocks with all those dinosaurs.

By Upper Paleocene times, mammals had begun to diversify, and to include some fairly large herbivorous beasts in their midst. These large, lumbering, downright ungainly mammals all turned out to be evolutionary dead-ends. Uintatheres, among the very latest of these archaic herbivores (lasting well into the Upper Eocene), are a prime example. Named for Utah's Uinta basin (just south of the Uinta range, the only range of the Rockies that runs east-west), uintatheres reached the size of rhinoceroses. Romer calls them "grotesque," a fairly colorful description in an otherwise rather staid scientific discourse. But the description is apt, nonetheless, as uintatheres were decked out with pairs of bony horn-like projections on their skulls. They also had expanded, fang-like canines that impart a rather menacing leer.

Another archaic mammal, *Coryphodon*, is fairly abundant in Lower Eocene rocks. Reaching a length of some eight feet, these herbivores belonged to a Paleocene-Eocene group known as "amblypods." Early occupants of the "large herbivore" niche in North America during early Tertiary times, they, too, have disappeared without a trace. In fact, they are extremely rough ecological equivalents of later herbivores—odd-toed perissodactyls (similar to today's horses, rhinos, and tapirs, but including as well the extinct and often huge titanotheres, or brontotheres), and the even-toed

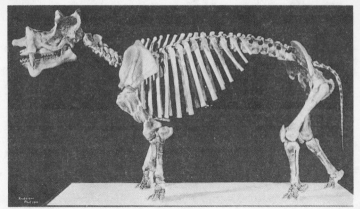

FIGURE 76   The Eocene *Uintatherium*.

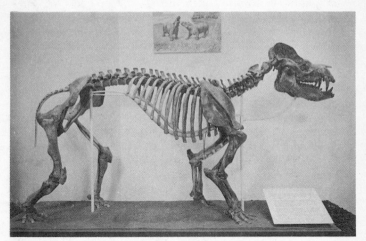

FIGURE 77   *Coryphodon*, another primitive mammal from the early Tertiary.

artiodactyls (pigs, camels, deer, antelope, and cattle, plus some extinct groups). Artiodactyls and the perissodactyls each invented a peculiar form of ankle joint—one that made it possible to be fleet of foot. The artiodactyl astragalus (ankle bone) literally offered a key to evolutionary success (paleontologist George Gaylord Simpson referred to such adaptations as "key innovations"). At the very least, it is the sort of structure that gladdens the paleontological heart: in one variant form or another, it is shared by absolutely all artiodactyls, and hence is evidence of the evolutionary integrity, the *monophyly*, of the mammalian Order Artiodactyla.

Horse Evolution: A Microcosm of the Mammalian Fossil Record

Artiodactyls and perissodactyls both got going in the Eocene. Horse evolution became an important focus of research almost immediately after Darwin published his *Origin*. Thomas Henry Huxley, the Darwinian "bulldog," came to the United States in 1876 to lecture and spread the evolutionary word. By that time, a number of fossil horse species had been discovered in Tertiary rocks of the American West. And there was some confusion: horses had been imported into North America by European explorers. Early Tertiary horse-like creatures were known in Europe: palaeotheres seemed to be perfectly plausible ancestors of European horses, though they last appear in the European fossil record in the Oligocene. Huxley also came to examine the American fossils—and saw the rudiments of a story that has grown in richness through the later efforts of many paleontologists, not least among them George Gaylord Simpson, who devoted an entire book to the subject of horse evolution.

Horses seemed to show a perfect picture of gradual evolutionary change through time. The earliest one is *Hyracotherium* (originally named *Eohippus*, or "dawn horse" in North America), known from Eocene rocks in both Europe and North America. The size of a smallish dog, and equipped with teeth very similar to those of the earliest rhino, *Homogalax* (a contemporary of *Hyracotherium*) *Hyracotherium* seemed the perfect ancestor to all later horses. As time went by horses became progressively larger. Their teeth grew longer (as they switched from browsing to grazing), and their faces grew longer. And the number of toes on both front and rear legs decreased—from four in the front and three in the rear in *Hyracotherium*, to one toe all around in modern horses.

These are some of the simple facts of horse evolution. Simpson pointed out, though, that horse evolution was much more complicated than a simple, straight-line picture would have it: there were several side-branches, for example, and not all genera alive at any one time were equally anatomically advanced. Still, the picture of horse evolution as a series of correlated trends in progressive anatomical change prevails, and is in basic accord with the facts. And herein lies a tale still very much relevant for modern science—and even for the public at large.

There is an exhibit on horse paleontology and evolution in the Hall of Tertiary mammals at the American Museum of Natural History. Lined up, from left to right, are skeletons of the Eocene *Hyracotherium* and *Orohippus*, the Oligocene *Mesohippus*, the Miocene *Merychippus*, and right up to Pleistocene *Equus*, the same genus as our modern horse. Sure enough, lined up as they are in their stratigraphic order of occurrence, they graphically display the various anatomical trends in the evolution of horses over the past 55-odd-million years. Undoubtedly one message that exhibit intends to convey is the gradual and even directional nature of horse evolution throughout that period. That is only natural: such men as William Diller Matthew and George Gaylord Simpson, Matthew's famous successor at the American Museum, believed that the historical pattern of evolutionary change was *gradual* and *progressive*. In this exhibit, they were merely displaying the fossils themselves,

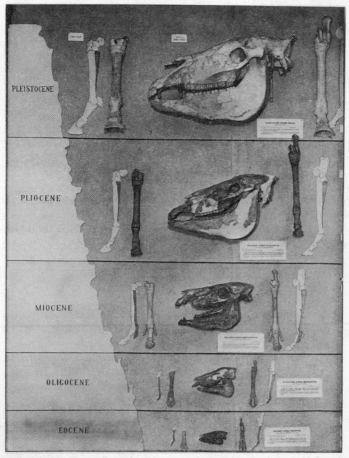

FIGURE 78 Part of the exhibit on horse evolution placed on display at the American Museum of Natural History in the 1920s.

letting their appearance and correct stratigraphic order speak for themselves. But the message comes through loud and clear: anyone looking at that exhibit is bound to come away with the notion that evolution is a matter of gradual, progressive change through time.

The exhibit merely shows one well preserved specimen of one species per unit of Tertiary geologic time; most of those species persisted for considerable spans of time and display the sort of anatomical stability I have come to expect as the rule in the fossil record. Horse evolution may well have occurred in the same fits-and-starts fashion that we have encountered in many other groups—and still be consistent with the old display that seems to imply progressive, gradual change through time. There are many ways to interpret what we see displayed in the exhibit cases. But, no matter what our evolutionary deductions, the anatomical facts remain the same.

Here is where larger, social issues can sometimes impinge on what would seem like purely scientific matters—judgments about exactly what the evolutionary significance of those horse fossils might be. When the creationist movement began to gather steam once again in the early 1980s, I was asked to appear on a network news show to explain the paleontological evidence of evolution. I chose to use that very horse exhibit to show how the fossil record—the actual sequence of fossils—very much agrees with the overall notion that life has had a history, and that the forms of related organisms indeed do change as we trace them through the rock record. Somewhat later, a critic of the modern evolutionary theory appeared on a NOVA show and actually made the erroneous claim that the fossils were arranged to show anatomical trends through time, i.e., were not displayed in their proper chronologic order—presumably an innocent, if somewhat unforgivable, error on his part.

Creationists often accuse paleontologists of arranging their data to fit their theories. But the problem in science is to see how those "facts," such as the data of horse evolution, fit various alternative theoretical notions of how life evolves. It is easy to see gradual evolution if only one horse specimen from each time interval is displayed. But when we realized that each specimen represents an entire species that was but one of several species alive at any one time, and when we further recognize that each species tended to remain stable for at least a million years, we get an alternative, more "punctuated" picture of evolutionary change. I would hate to see that lovely old exhibit dismantled—it is a classic of paleontologic museology, and the specimens, as always, are simply terrific. But we should all be aware that evolutionary patterns are seldom displayed to their fullest when only a few specimens can be placed on exhibit.

The horse story simply shows that what constitutes honest scientific debate is often misconstrued outside of the relatively narrow confines of a scientific discipline: the nature of the dispute may not be thoroughly grasped—or far worse, the mere fact that there is disagreement is often used as evidence that scientists really do not know what the score is, and are perhaps even guilty of arranging the evidence to fit their preconceptions about the way the world really *ought* to be. Scientists, being human, are often prone to admit that such undetected biases can creep in, but they also know it is their job (and, especially, their colleagues' job) to make sure they keep on a straight and narrow path of objectivity. Creationists would have us believe that actual chicanery, finagling the evidence, is used to support the idea of evolution. Correct the errors and the very evidence for evolution itself goes away—or so they think.

Right now, creationism once again seems on the wane—although history has shown us that it never goes away completely. At any rate, it is absolutely amazing what those horses have prompted people to say about evolution.

Marine Rebounding

Meanwhile, back below the waves, life in the early Tertiary was also busy springing back from the Cretaceous collapse. Gone were the ammonites. In their place, early in the Tertiary, there was actually a modest radiation of that other group of externally coiled cephalopods, the nautiloids. As we have seen, the nautiloids had been abundant and varied in the Lower Paleozoic, especially in the Ordovician, but also in the Silurian, but their diversity had nosedived as the ammonites appeared and quickly diversified in the Middle and Upper Devonian. There were nautiloids in the Upper Cretaceous—*Eutrephoceras* was a somewhat globular version of the modern *Nautilus*. But in the Paleocene and Eocene the nautiloids assumed a variety of shell shapes and even developed a sinuosity to their suture patterns that is reminiscent of the relatively simple patterns of the Paleozoic goniatite ammonoids.

Much the same thing happened in the Lower Carboniferous. All trilobites except proetids had become extinct by the end of the Devonian. And sure enough, the proetids underwent a modest little radiation of their own in the Lower Carboniferous, very much as if they were reacting to vacancies created by the demise of other trilobites.

But care is required in such speculations. After reading the last two paragraphs carefully, one might be tempted to conclude that large groups (*taxa*) are actual units that compete with one another. And that would be tantamount to calling them economic entities, capable of interacting in the ecological arena of life. But there is no way in which ammonites, say, literally can compete as a unit, a single entity, with anything, including their nautiloid cousins.

Yet the patterns are definitely there. Their best interpretation, in my opinion, is that when extinction claims a species, or many species, the most likely candidate to fill a similar ecological niche as life bounces back would come from the ranks of the closest remaining relatives. Note that this doesn't always happen: mammals are synapsid amniotes, while dinosaurs were diapsid amniotes. Birds, also diapsids, did radiate in the Tertiary, but the vacancies created when the dinosaurs went were taken by mammals, not birds, or, perhaps more significantly, lizards and snakes. Mammals, seemingly, were more similar to dinosaurs already in terms of their basic modes of economic existence. So it fell to them to take over.

But note, too, that the hierarchically arranged structure that links together all of life means that everything is related to everything else: it is just a question of degree. Granted that mammals are less closely related to dinosaurs than are, say, birds or crocodiles, it is still noteworthy that they and dinosaurs are both amniote vertebrates: dinosaurs were not succeeded in any meaningful way by a radiation of insects, mollusks, or any of the other (relatively few) groups of animals that have managed to become established on land.

But there is still another theme in the marine realm: if the Permo-Triassic witnessed a greater disruption in the sea than it did on land (at least in some parts of the globe), the situation seems rather the reverse for the Cretaceous-Tertiary extinction event. True, the ammonoids went, as well as the rudists (huge, coral-shaped clams that clustered to form reefs) and a number of other groups. But graphs of diversity of marine bivalves, snails, and teleost fish quickly reveal an unbroken rise beginning in the Upper Jurassic or Lower Cretaceous—an expansion in the number of families (hence, presumably, of species) that is fairly regular and steady for all three groups—and that seems virtually unaffected by the great Cretaceous extinction event. The extinction was severe; many species died out. Marine micropaleontologist M. N. Bramlette's graphs for coccolithophorid species is especially shocking: some twenty-seven species simply drop out at the end of the Cretaceous, to be replaced by many new species in the Lower Paleocene. Much the same went on for virtually all marine groups; but large-scale units, such as clams, or snails, or teleost fish, did not suffer extinction. Consequently, in the marine realm, little in the way of the radically new appeared after this particular extinction event. The moral is clear:

the greater and more far-reaching an extinction event—i.e., the higher the level of the groups that are knocked out—the greater the probability that new, equally highly ranked groups will appear. They may come as ecological analogues, or they may represent something more nearly economically "original," but in any case they will be seen as evolutionary novelties on a relatively grander scale than the new creatures that appear after a less all-encompassing extinction.

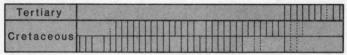

FIGURE 79 Diagram showing how many species (vertical lines) of calcareous algae became extinct right at the Cretaceous/Tertiary boundary, how many made it through, and how many appeared for the first time in the Lower Tertiary, in two widely separated regions. From Bramlette, M.N., in *Science*, Vol. 148 (1965), pp. 1696-99, Fig. 1. Copyright 1965 by the American Association for the Advancement of Science. By Permission.

There is something about a visit to a richly fossiliferous Tertiary marine outcrop that is awfully familiar. The sediments tend to be unconsolidated (though I have seen some ·mighty hard Tertiary rocks)—and collecting fossils from loose sand is a lot easier than chopping them out of rocks. The effect of the sand itself is marked: it is sometimes hard to tell whether you are collecting fossils or, especially if you happen to be standing somewhere along the coastline, whether you are collecting modern clams and snails from the beach. There are places in Maryland, Virginia and on down the Atlantic Coastal Plain, where the fossils are so fresh, so beautifully preserved, that they hardly seem like fossils at all—actually a source of some disappointment to ardent collectors trying to find a "fossilized" remnant of exotic marine life. Usually, the fossils are a pure white, but sometimes even traces of the original color markings can be seen—if not by the unaided eye, then by exposing the shells to ultraviolet light, which may cause old pigment spots to fluoresce slightly.

This gets us back to Charles Lyell and his measures of "recency" of Tertiary molluscan assemblages. It takes someone more than just casually familiar with clams and snails to know that Eocene or Miocene species are at a particular locality. And that is because the Cretaceous extinction failed to knock out the major groups—snails, clams, and teleost fish—that were already expanding, filling the seas with increasing numbers of species. By Cretaceous times, the snails and some other major groups were already beginning to look "modern."

The Gastropod Mollusks

Snails, in particular, impress one as essentially modern as far back as the Cretaceous. Snails (gastropod mollusks) are traditionally divided into prosobranchs (the bulk of marine snails), opisthobranchs (which include groups that have lost their shells, particularly the marine nudibranchs) and the pulmonates, opisthobranch derivatives that are our land snails and shell-less offshoots, the slugs. (The mascot of the University of California at Santa Cruz is not a macho mountain lion, or anything of the kind: it is a slug, namely the "banana slug," a huge, yellow-green number that occurs in fairly high numbers around the campus. It may be the only invertebrate mascot in the land.)

Marine prosobranchs—the vast majority of snails that find their way into the fossil record—are also divided (like Gaul and so many other things in this life) into three. Archeogastropods are the most primitive of the three groups. In the Paleozoic, they included the bellerophontids and the abundant and varied pleurotomarians, descendants of which (including abalones) still inhabit the seas today, albeit, like the brachiopods, in reduced numbers. Trochids, also archeogastropods, have been common from Paleozoic times right up to the present. Mesogastropods, the second group, likewise predominantly herbivorous, have some advanced internal features. First appearing in the Ordovician, they include a rather large variety of groups. Archeogastropods and mesogastropods, as their names virtually imply, are *grades* of anatomical development, rather like the goniatite-ceratite-ammonite division of ammonoids. But unlike the ammonoid situation, the snail grades are probably not evolutionarily unified groups: some archeogastropods are probably more closely related to some mesogastropods than they are to other archeogastropods. In any case, it is not until the appearance of the neogastropods in the Cretaceous that we seem to be dealing with a coherent, "monophyletic" group of snails. Neogastropods are carnivorous. All have siphons for the expulsion of waste products from the bottom of the aperture (shell opening). Many have

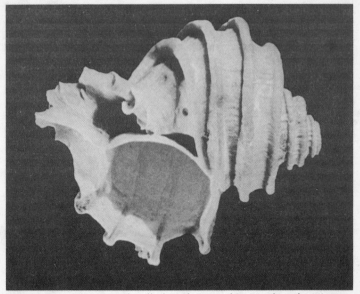

FIGURE 80 The gastropod mollusk *Ecphora quadricostata* from the Miocene of eastern North America. This was the first North American fossil species to be described in the scientific literature.

siphons, as well, for the intake of fresh water near the top of the aperture.

Neogastropods include such familiar snails as the cones (deadly poisonous), whelks, and olives. More even than insects, they are the object of serious study and collection by amateurs, a practice that also encourages familiarity with their fossil record. Indeed, groups such as neogastropods and the various modern families of bivalves are perhaps the most promising creatures to tackle in the fossil record. They are still very much alive today, so that we can come to grips with the biology of living creatures. And neogastropods fossilize very well, allowing us to get an accurate reading of their evolutionary history through nearly 100 million years of geologic time.

Evolutionary Rates and the Mid-Tertiary

In 1944, vertebrate paleontologist and mammal specialist George Gaylord Simpson published *Tempo and Mode in Evolution*, a book that was designed to bring paleontology in line with the newly emerging "Modern Synthesis" of evolutionary theory. A great success, *Tempo and Mode* established Simpson as the leading paleontological student of evolution, a position he held until his death in 1984.

Part of Simpson's approach to the welding of genetics with fossils came in his work on evolutionary rates. Indeed, in his preface, he tells us that "how fast do animals and plants evolve?" is the very first question a geneticist will ask of a paleontologist. And for good reason: as Simpson accurately foresaw, how fast we think creatures evolve determines in large measure *how* we think they evolve.

Simpson developed a moderately complex classification of kinds of evolutionary rates. At base, he wanted to know how fast the genetic information changes—and to measure that with fossils, one would have to measure the rates of change in size and shape of anatomical structures of organisms. Indeed, until protein and (later) DNA sequencing came along (telling us the actual composition of the genetic material), even geneticists had to rely on the external features of organisms to study their genes. (Remember Gregor Mendel, with his smooth and wrinkled peas.)

Because comparing fossils collected at different geological horizons to measure evolutionary rates is filled with practical difficulties, Simpson suggested a shortcut: we can simply record the rate that new species, or even genera, appear and disappear in the various subdivisions of geologic time. Simpson assumed that classification of organisms into species and genera is a reflection of the amount of genetic change that has occurred: it takes more genetic change to produce a new genus than it does to produce a new species—a risky but not wholly unreasonable working hypothesis.

Simpson drew on his extensive familiarity with fossil mammals, calculating the characteristic rates of appearance of new families per unit of geologic time. (Ideally he would have used species, or genera, but we have already seen how families are easier to sample, and thus in a way the data are more reliable.) For comparison, he turned to the literature on fossil bivalved mollusks. And what he found was perhaps surprising: mammals seem to evolve many times faster than bivalves! Even if we might expect some disparity between characteristic rates of evolution in different groups, the contrast Simpson found between clams and mammals seemed a bit excessive, prompting evolutionists to offer explanations right down to the present day.

One possible explanation of the clam-mammal disparity in evolutionary rates is that the phenomenon is not "real"—there is something wrong with the data, or our analysis of it. It is quite possible, for example, that techniques and traditions may have been different among clam and mammal students; in particular, what would be regarded as a genus among clams may be classified as a family among mammals, rendering the levels not truly comparable between groups. In other words, the taxonomic system itself may artificially suggest a difference in rates where in fact none really exists.

A modern version of this objection was published by paleontologists Thomas J. M. Schopf, Stephen Jay Gould, and David M. Raup, and biologist D. Simberloff. They argued that organisms with more complex fossilizable anatomies are bound to show faster rates of evolutionary change since there will be more for the taxonomist to go on: evolution is easier to spot in more complex forms. And mammal skeletons, one might suppose, are vastly more complex than clamshells.

Perhaps—though in fact there are many ins and outs, many guides to the complicated soft anatomy, to be found on a clamshell. If we start from another vantage point, we see that these objections carry no real, overwhelming authority. If we assume that species actually exist in the real world, it is the paleontologist's job to find them, using whatever anatomical clues filter through the vagaries of fossilization, exploration, and collection. As complex as a mammal's skeleton is, most classificatory work on fossil mammals is done on their teeth—typically complex affairs, but certainly no more so than an entire clamshell. And even though there may be different canons of technique traditionally employed by clam and mammalian paleontologists, the more rapid rise and fall of entire families seems, at bottom, really to reflect a higher rate of speciation in mammals than in clams. In other words, the disparity in evolutionary rates that Simpson thought he detected seems to be a real phenomenon, requiring an explanation in evolutionary terms.

I believe that there is nothing intrinsic about "clamness" or "mammalness" that either confers characteristic rates for each group or accounts for the differences between them. Rates of evolution vary enormously *within* both clams and mammals—there really is no meaningful "characteristic" rate. What differs between the groups is simply the *average* evolutionary rates. And the explanation seems to be that *terrestrial* organisms in general evolve at higher rates, regardless of how those rates are measured, than do creatures that live in the sea. And the reason for *that* seems to be that the world is a lot more heterogeneous and changeable under a blanket of air than submerged in the briny deep.

From the standpoint of adaptation alone, the diversity of physical circumstances on land would be expected to result in a proportionally greater array of anatomies than we might expect in the sea. But something more is afoot: terrestrial environments seem much more prone to change, either prompting evolutionary change or bringing on the outright demise of species, than aquatic environments. (This excludes the truly mass extinctions, which can be equally devastating on land and in the sea.) Speciation, the origin of new reproductive communities from old ones, is mostly a matter of geographic fragmentation in which environmental heterogeneity plays a leading role. Not just mammals, but all land-living creatures show higher average group evolutionary rates than what we typically see in the marine situation.

Clam species, like most other marine invertebrates, tend to last from 5 to 10 million years, with exceptions, of course. Some live a lot longer; others suffer extinction much sooner. Mammal species tend to last only 1 or 2 million years, again with plenty of exceptions. The higher rate of speciation and extinction in mammals, then, should lead to a much greater total amount of anatomical (and genetic) change per unit of geological time. However, though I fully admit that such a comparison of total amount of anatomical change between two groups with no anatomical features in common (save in their cellular anatomy and biochemistry) is impossible to measure in any precise way, the impression one gets is that some mammals may be just as prone as clams to getting stuck in evolutionary "ruts"—to keep churning out new species that look pretty much alike.

There is a parallel example mentioned in Simpson's *Tempo and Mode*. One of the factors in evolution, he mused, must be generation time—the length of time it takes for organisms to grow up, mature, and begin producing progeny of their own. Since natural selection works by biasing the reproductive success ratio of organisms within a population, the quicker the generation time, the more opportunities for selection to work—hence the faster the rates of evolutionary change ought to be. But Simpson realized that elephants, with their slow generation time, have evolved many times more quickly than oppossums, which have much shorter generation times. In other words, we can observe rapid turnover, yet the system as a whole remains the same.

Mammalian evolution, ever since the Eocene at least, seems to have settled down *in general*, to what Joel Cracraft of the University of Illinois and I termed a "steady-state" situation (shades of Charles Lyell!). True, horses continued to become modified in a complex directional manner. And our own remarkable lineage continued to change until we ourselves appeared comparatively recently on the geological-evolutionary scene.

Yet by Miocene times, most of the modern groups of mammals, the *families*, had already appeared. Little by way of the radically new has come along since. Species come and go, at rates much higher than the relaxed pace we see in the sea. Nor are all these new species in any sense mere "clones" of their ancestors. But for the most part we see variations on basic themes. Indeed, biologists of generations past sometimes thought that higher taxonomic groups, such as orders and families, characteristically appear first, followed by genera, then species. The reason: mammals, as a class, go back to the Triassic. But few, if any, living species of mammal go back much beyond the Miocene. That's the pattern, but of course the earliest mammalian organisms belonged every bit as much to a species as they did to a family, order, or Class Mammalia. What the pattern really means in evolutionary terms is that anatomical innovations—the features, such as hair, mammary glands, and three bones in the middle ear, that stamp groups such as the Class Mammalia—occasionally evolve, creating opportunities and setting limits for further evolutionary modification. Thus, we commonly see themes followed upon by variations. And that is what much of mammalian evolution was like in Middle and Upper Tertiary times.

The moral of this little analysis of evolutionary rates is actually quite profound; indeed, it promises to precipitate a fairly radical revision of evolutionary theory. For if molecules sometimes change rapidly inside the genetic machinery of the cell, but without causing much modification of organisms (a phenomenon recently documented by molecular biologists); if generation times are not correlated particularly closely with rates of anatomical change; and if rapid rates of speciation frequently seem to lead nowhere in

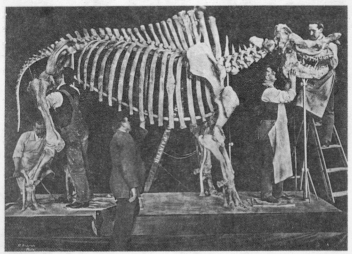

FIGURE 81 *Brontotherium*, a large mid-Tertiary relative of horses, rhinos, and tapirs.

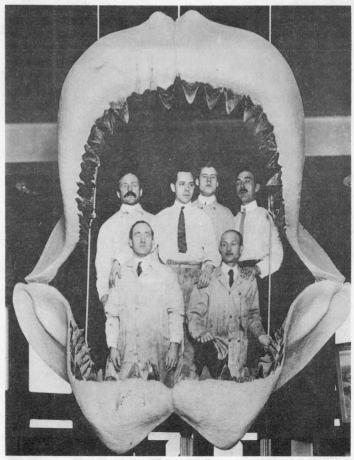

FIGURE 82 Jaws. Huge Miocene teeth of *Carcharodon megalodon*, larger relative of today's Great White Shark, inspired this reconstruction, which still hangs at the American Museum of Natural History in New York.

particular in terms of evolutionary change—evolution emerges as a multilevel, or *hierarchical*, affair. What goes on at one level may have little effect on the next higher or lower level. The comings and goings of entire species—and even larger groups—may have little to do with the normal processes of genetic change that go on from generation to generation within species. Paleontology has been burdened with the tast of understanding all evolutionary patterns in terms of genetics—as might be imagined, a virtual impossibility. Ironically, the situation stems in large measure from the success of

Simpson's book, as he strove to show that paleontological data were indeed consistent with what was then known about the genetics of the evolutionary process. But now we see that species have births and deaths, just as organisms do, but with different rules governing the process. The truly large-scale events, the sorts of things paleontologists are privy to, are to some degree independent of the sorts of genetical processes best studied with experimental organisms in the laboratory. This has the effect of bringing paleontologists back into the game to play an active role in evolutionary theory. And better yet, by promising a better fit between our ideas of how evolution occurs and the actual events in the history of life, we are bound to come up with a better evolutionary theory than we have managed to produce so far.

Onward and Upward: Plio-Pleistocene Times and Human Evolution

Sociobiology has been all the rage these last few years. Controversial, tending to go overboard in explaining all manner of human behaviors as manifestations of underlying genetic makeup, sociobiology nonetheless emphasizes one cardinal fact of nature—a fact that has had a curious history in human thought: mankind, genus *Homo*, species *sapiens*, is an *animal*.

Sociologist Kenneth Bock, in his *Human Nature and History* (1980), has reminded us of the obvious: that mankind is a species of animal life was known to the Greeks, the Egyptians, and, no doubt, to the earliest of our kin who were capable of giving the matter any thought. But it has long been a tradition for us to distinguish ourselves as a *peculiar sort* of animal. Our physical and behavioral uniqueness that has by now enabled us to overrun the planet in very much the manner of weeds and "pests" is often exalted as some pinnacle of evolutionary perfection. And viewing ourselves as the crown jewels in the diadem of evolutionary creation fits in nicely with the older conceit that we were created in God's own image. So we usually find discussions of human evolution tucked away as the final item in books that tackle the entire history of life.

And, actually, it makes a lot of sense. After all, hominids became upright, and the particular lineage that has led to ourselves became clearly differentiated, only a few million years ago. We are a relatively recent phenomenon. We are primates, though, and primates in general retain a large number of primitive mammalian traits. Indeed, our five fingers and toes are primitive early tetrapod features. Primates appear fairly early in mammalian evolution—back in the

Cretaceous, as a matter of fact, long before artiodactyls, perissodactyls, elephants, bats, and most of the other major groups of mammals. Alfred Romer, the great vertebrate paleontologist and comparative anatomist, actually positioned his chapter on primates (with its review of the fossil record of human evolution) near the beginning of his section on mammals in his *Vertebrate Paleontology*. It was a decision that made sense on the rational grounds of our true place in mammalian nature (primates on the whole are rather primitive mammals)—but still seems out of place to me.

Whatever our personal beliefs about the nature of mankind may be, we are nonetheless undoubtedly animals, and we have evolved in every bit the same sense as have all other organisms. Thus it comes as no great surprise that our lineage actually has a moderately good fossil record. An unprecedented exhibition, in which many of the most important of these fossils were gathered together at the American Museum of Natural History in 1984 (an event unlikely to be repeated soon, if ever), laid this evidence out before the eyes of all who would look. The anatomical connections between our fully modern selves and our remote australopithecine ancestors of the African savannahs of 4 million years ago were compellingly obvious.

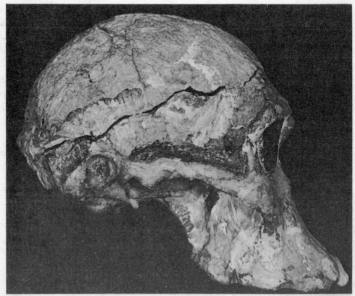

FIGURE 83 "Mrs. Ples"—skull of *Australopithecus africanus* from the Lower Pleistocene of Sterkfontein, South Africa—one of the earliest members of the human family tree.

We have indeed evolved; in particular, our brains have become relatively larger as our overall body size has increased. And all of this has happened pretty quickly: it took but 4 million years for our brains to expand from approximately the size of Lucy's ("Lucy" being Donald Johanson's most famous specimen of the species *Australopithecus afarensis*) up to the biggest brained of all hominid species: *not* ourselves, but neanderthals, who lived in Europe from roughly 100,000 to 35,000 years ago.

There has been a bit of collision in recent years between applied molecular biology and human paleontology. Paleontologists for years have assumed that the earliest australopithecines (for many years known only from southern Africa) were perhaps only 2 million years old. Obviously, they thought, there was a long gap between australopithecines and the line that led to the great apes. *Ramapithecus*, known from both Asia and Africa, seemed to some paleontologists to represent an early form of hominid. Its age was variously estimated to range from about 18 down to 8 or 9 million years ago.

Then came the notion of molecular clocks: assuming (as seems reasonable at least for many cases) that mutations that account for minor differences in the composition of protein molecules accumulate at a statistically almost constant rate, we can measure how similar species are to one another. And if we know the rate of those mutations, we can estimate the time of divergence of the lines leading to the now separate species. Finding that we share 99 percent genetic similarity with chimpanzees (a rather mind-boggling statistic), molecular biologists calculated that our line split off from the chimp lineage only about 4 to 5 million years ago. The discrepancy between the dates was significant: paleontologists were calling for a much earlier divergence of man and ape lines than the molecular data seemed to support.

And the molecular data now seem much the more accurate. But the last word has not been written: for example, paleontologist and anthropologist Jeffrey Schwartz of the University of Pittsburgh has recently argued, using a wide range of criteria, that our closest relatives are orangutans, and that *Ramapithecus* shares a number of features with orangs. Orangs have further specialized; many of our similarities with chimps (and gorillas) could well be primitive retention of old hominid features, not especially indicative of a relatively recent divergence. Often construed as a triumph of modern science over the fusty old study of moldering bones, molecular techniques of evolutionary analysis are as prone to misinterpretation as are those old bones. But there is a positive side to all this: the more data we have, the more likely we are to come up with an accurate picture of nature. Discrepancies are actually the very essence of science: if we don't turn up anomalies that must somehow be reconciled with our ideas and observation of nature, we will continue to think we have gotten the story right.

Why the difficulty in interpreting these old bones? It is because, with the exception of a few very clear-cut specializations (such as arms modified for swinging through trees in the great apes; upright stance and expanded brains in ourselves), we have remained conservative mammals. Especially in our own lineage over the past 4 million years, it is clear that we have sprung from a conservative line. There have been offshoots of rather more specialized species, such as the "robust" australopithecines of Africa, with their massive jaws and big teeth that stamp them as herbivores. We are from the "gracile" line. Our diet today, diverse as it is for most of us taken as individuals, is fantastically varied if the whole range of human food consumption is considered. Our culture—the organized, learned systems of behavior in which we all, as humans, participate—is our greatest tool of adaptive equipment. It is culture, more than anything else, that has enabled Homo sapiens to range so widely over the Earth's surface.

Thus our overall profile is that of an ecologically generalized, flexible species. And such a strategy usually implies evolutionary conservatism: there is little need for anatomical transformation, the sorts of evolutionary inventiveness encountered in species that become specialized. We are specialized in one important respect: our cultural capacities, reflected, of course, in our large and inordinately complex brains. Ironically, it is this extreme specialization that has enabled us otherwise to remain quite generalized. And therein lies the explanation for the difficulties in sorting out the early fossils leading to apes and humans: when specialized anatomical features are present, it is easy to see who is related to whom. Hair unites all mammals. The main trunk of hominid evolution simply has not accumulated very many anatomical specializations at all. Rather like the (as yet unresolved) case of the molecules themselves, retention of primitive features confuses the picture, making it difficult if not impossible to estimate when species diverged simply because everything continues to look pretty much alike.

Is human evolution gradual or punctuated? Is there a nice steady stream of gradual modification, especially increase in brain size? Or are there long periods of stability, "punctuated" by occasional events when new species arise? The answer is "yes" to both questions, and the reason in all likelihood is that we are such a mixture of generalist and specialist. (We are, in a way, a specialized generalist species.) There do seem to be long periods of stability, even within our own direct ancestors. *Homo erectus*, the Middle Pleistocene species of Europe, Asia, and Africa, seems to have remained about the same in overall brain size (and cultural adaptation, as judged by the tools left behind), though it comes as no surprise that this interpretation is the focus of debate. Sudden speciation and consequent anatomical stability, though, seem more characteristic of the species of robust australopithecines, out on the specialized, herbivorous side branch of the human tree.

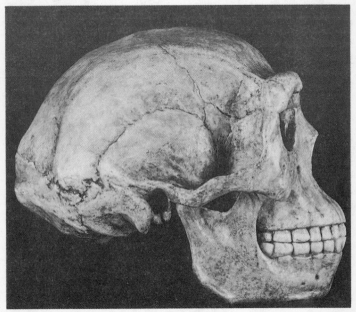

FIGURE 84   A reconstruction of *Homo erectus*, a mid-Pleistocene forerunner of *Homo sapiens*, from Choukoutien, near Beijing, China.

Homo sapiens in anatomically modern form showed up on the European scene about 35,000 years ago, just when neanderthals disappear from the record. And even here we have debate, with some paleoanthropologists insisting that neanderthals quickly transformed into ourselves in proper linear fashion. But there is evidence from the Mideast and Africa that suggests that a transition to modern humans was already underway 50,000 years ago (and perhaps even earlier). It looks very much as if our immediate progenitors evolved elsewhere and simply invaded Europe. The radical change in cultures, with the advent of cave paintings perhaps the most striking element, seems to support this view of migration, with the disappearance of neanderthals seen as a true extinction event. I need hardly point out that such scenarios agree well with much of human history since then.

Pleistocene Extinctions, the Evolutionary Here and Now, and a Few Thoughts on What May Be in Store

We are high and dry on the continents today simply because the Earth's polar ice caps have locked up so much water. Normally, as we have seen, the continents are inundated. The four or five major pulses of glaciation in the Northern Hemisphere brought huge continental ice sheets migrating southward over much of Europe, Asia, and North America. An ice sheet still covers most of Greenland. Naturally all this glacial activity played merry hell with the climate of the entire world.

Paleontologist Elisabeth S. Vrba of Yale University has noted a big evolutionary burst in Africa occurring just over 2 million years ago, apparently a reflection of a major ecological change in the vegetation, itself correlated with a cooling episode associated with a major pulse of glaciation. The Plio-Pleistocene is a wonderful source of examples of extinctions and proliferations, all as a response to wholesale environmental change. It is a mini-version of the major themes in the history of life, and, consequently, of this book.

But the extinctions that have gone on in recent years have taken on an additional aspect. Paleontologist Paul Martin of the University of Arizona, in particular, has sought to implicate the loss of large mammals over much of the world with the advent of hunting peoples. There are some obvious problems: many small mammals have become extinct as well, and in Africa, where human hunting has gone on at least as long as anywhere else on Earth, though there were Pleistocene extinctions aplenty, the net effect on the mammal fauna was by no means so severe. Africa, especially eastern and southern Africa, really is a huge menagerie of Pleistocene mammalian life, providing a wonderful glimpse into the past as well as preserving an imperiled treasure of the Recent.

Yet there is no doubt more than a grain of truth to Martin's thesis that extinction of large mammals was greatly exacerbated by mankind, just as mankind continues to be the gravest threat to the future existence of all those species in Africa—and everywhere else. At first glance, inclusion of mankind as part of the cause of major extinctions seems not to fit with the general picture we have of other, more remote extinction events. But this objection is put to rest when we remember that we, too, are organisms, a species of mammal with local populations spread out over almost the entire surface of the Earth and playing a direct, economic role in nature. Our local populations everywhere have niches every bit as much as the local populations of all those other species. Indeed, our putative role in Pleistocene extinctions, a role that is obviously still very much being played as the world loses species at the apparent rate of one a day, strengthens my conviction that mass extinctions take many thousands of years, and reflect both earthly physical environmental *and* biological causes. In other words, even granting mankind's probable role as active agent, the Pleistocene extinctions, which have taken such exotic creatures as mammoths from North America as recently as 10,000 years ago (*think* of that!) are very much the same sort of affair as all those other large-scale extinctions we have seen in the geological record.

Can we expect, then, a reproliferation? Probably yes, after a lag period. But we are still very much in the extinction mode. And we

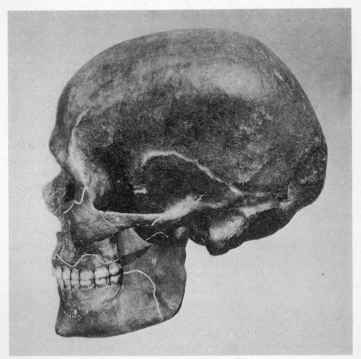

FIGURE 85 Cro-Magnon man, a fossilized member of our own species, *Homo sapiens*. We must endeavor to avoid extinction, for only if we survive can we continue to contribute our own bones to the fossil record!

can do something about *that*—as we had better do. It is one thing to take a dispassionate look at life's history, to observe that extinction is normal, and that it paves the way for the truly new. It is another matter indeed when we come to consider our own position—as one of the many species currently occupying mother Earth. Extinctions *are* real. We could easily go, too—as scenarios of "nuclear winter" have recently told us. (The parallel between extraterrestrial impact scenarios, with their clouds of particles occluding the sun and shutting down photosynthesis, and "nuclear winter," with hydrogen weapons producing the same effect, has been drawn. Notwithstanding my position on the extraterrestrial impact scenario itself, I remain convinced that nuclear warfare is bad for the health of all organisms.) More to the point is the older view of the "ecology movement": we really are integrated into the ecosystems of the world, and the loss of too many of our fellow species itself is a grave threat to our own continued existence.

Thus extinctions, in the abstract consideration of life's entire history, have paved the way for the truly new, including, of course, ourselves. But now that we are here, it is not inconsistent at all to want to see our own species survive. To hell with innovation, let's try to stick around!

Glossary

adaptation   a structure, physiological process, or aspect of behavior of an organism shaped by *natural selection* to perform a specific role or task; also, the process of developing such features in *evolution*.

biomere   *a biostratigraphic unit* of *stage* magnitude containing a coherent evolutionary fauna, with upper and lower boundaries that may transect time; first described by A. R. Palmer.

biostratigraphic unit   any body of rock delineated by its fossil content, including *zones*, *stages*, and *biomeres*.

biostratigraphy   study of the distribution of fossils in sedimentary rocks, especially the use of fossils in temporal *correlation* studies.

continental drift   a theory, generally attributed to A. Wegener, that the continents have changed relative position over the Earth during geological history; forerunner of the modern theory of *plate tectonics*.

correlation   in geology, the demonstration that two or more bodies of rock share some equivalence, usually in terms of age of formation. *Fossils* play a major role in the correlation of *sedimentary rocks*.

DNA   deoxyribonucleic acid; the macromolecule housing the genetic information of nearly all organisms.

ecosystem   regional configuration of all organisms and physical environment; generally composed of populations of many different *species*, bound together by the flow of energy between the physical and organic components and between organisms.

epoch   smallest conventionally recognized subdivision of *geologic time*; subdivision of a *period*.

era   largest conventionally recognized subdivision of *geologic time*; divided into *periods*.

eukaryotes   organisms possessing an advanced type of cell, with the nucleus set off from the remainder of the cell by a distinct membrane; bacteria are prokaryotic (see *prokaryotes*), and all other organisms (except viruses) are eukaryotic.

evolution   the idea that all organisms, living and fossil, are descended from a single common ancestor; thus evolution is the genealogical history of life; also, the history of genetic information conserved, modified, and transmitted by organisms; also, the processes involved in forming such history.

extinction   termination of a *species*, or of an entire lineage.

formation   thinnest body of rock (generally *sedimentary*) that can be conveniently defined and depicted on a geological map.

fossil   any trace or remains of ancient life; generally reserved for specimens over 6,000 years old ("prehistoric"); includes actual body parts preserved chemically unchanged, as well as altered parts, imprints, and traces of the activities of organisms, such as burrows.

fossil record   the total sequence of fossils preserved in *sedimentary rocks*.

genus (pl., genera)   a category of the *Linnaean hierarchy* ranked below family and above *species*; also, a *taxon* that is a part of a family and that includes one or more species; the name of the genus, always capitalized, is the first of the two latinized names given to each species, i.e., "Homo sapiens"—we are members of the genus "Homo," species "sapiens."

geologic column   locally, the sequence of rock exposed in an area; globally, the composite sequence of rocks throughout *geologic time*.

geologic time   the period of existence of planet Earth, beginning approximately 4.55 billion years ago.

geology   the science of the Earth, including its history.

igneous rock   any rock formed by the cooling of a molten mass. Granite and volcanic rocks (e.g., basalts) are igneous rocks.

invertebrates   all complex animals (*metazoans*) that lack a vertebral column (spine), hence nonvertebrates; invertebrates constitute an evolutionarily heterogeneous array of animals.

Linnaean hierarchy   a system of categories used in the classification of all forms of life; ranging from kingdoms (e.g., Plantae, Animalia, Fungi) down to *species*, the smallest generally recognized taxonomic division of life. Named for Carl von Linné (Carolus Linnaeus), "father" of systematics or taxonomy i.e., the science of classification.

metamorphic rock   any rock formed by the application of heat or pressure to any other pre-existing rock. Slate, schist, marble, and gneiss are metamorphic rocks.

metazoans   true multicellular animals.

natural selection relative reproductive success of some organisms over others in a population, thus a determinant of the genetic composition of the next generation; reflecting relative success in the economic sphere (as in obtaining food or avoiding predators), natural selection maintains and modifies *adaptations*.

niche, ecological the economic role played by a population of organisms, all of the same species, within a local *ecosystem*.

paleontology the scientific study of fossils.

period a division of *geologic time*; periods are subdivisions of *eras* and are composed of *epochs*.

plate tectonics modern theory of the Earth's construction and history; the Earth's crust is composed of some eight major divisions ("plates") and numerous minor plates; the boundaries between such plates may be passive, or plates may slip past one another, generally causing earthquakes; also, crustal material may be created (as at the midocean ridges) or destroyed ("subducted") at plate margins.

prokaryotes organisms possessing a primitive cell type, lacking the advanced structures of *eukaryotes*, such as a discrete, membrane-bounded cell nucleus; all prokaryotes are bacteria.

punctuated equilibria term employed by N. Eldredge and S. J. Gould for a pattern of evolution consisting of long periods of stability ("stasis"), interrupted occasionally by periods of relatively rapid evolutionary change, interpreted as true events of *speciation*; contrasts with a pattern of long-term gradual, progressive change.

RNA ribonucleic acid, a class of macromolecule providing the interface between DNA and the manufacture of proteins; in many viruses, RNA (rather than DNA) is the molecule of heredity.

sedimentary rock rock formed from hardened (indurated) deposits of mineral particles; shales, siltstones, sandstones, and limestones are sedimentary rocks.

series a body of rock corresponding to an *epoch*, a division of *geologic time*; series are subdivisions of *systems* and are composed of *stages*.

speciation the origin of a new *species* (reproductive community), generally by the splitting off of a descendant from a parental species.

species a genealogical entity whose component organisms mate and reproduce among themselves but not (as the overwhelming rule) with organisms of other such entities.

stage a body of rock corresponding to an age, a subdivision of *geologic time*; stages are subdivisions of *series*, and are composed of one or more *zones*; see *biomere*.

system a body of rock corresponding to a *period*, a subdivision of *geologic time*; systems are composed of *series*.

taxon (pl., taxa) any genealogical entity composed of one or more *species* and corresponding to a ranked category of the *Linnaean hierarchy*; kingdom Animalia, class Trilobita, and species *Phacops rana* are all taxa.

vertebrates subphylum of phylum Chordata; animals with backbones, including various classes of fishes, amphibians, reptiles, birds, and mammals.

zone smallest conventionally recognized *biostratigraphic unit*; subdivision of both *stages* and *biomeres*.

Index